DINNER

at the

NEW GENE CAFÉ

DINNER

at the

NEW GENE CAFÉ

How Genetic Engineering Is Changing
What We Eat, How We Live,
and the Global Politics of Food

BILL LAMBRECHT

Thomas Dunne Books
St. Martin's Griffin
New York

FOR SANDRA OLIVETTI MARTIN,

MY BOSSY WIFE AND EDITOR

THOMAS DUNNE BOOKS.
An imprint of St. Martin's Press.

www.stmartins.com

Design by Victoria Kuskowski

Library of Congress Cataloging-in-Publication Data
Lambrecht, Bill.
    Dinner at the new gene café : how genetic engineering
is changing what we eat, how we live, and the global
politics of food / Bill Lambrecht.
        p. cm.
    Includes index.
    ISBN 0-312-26575-1 (hc)
    ISBN 0-312-30263-0 (pbk)
    1. Genetically modified food. I. Title.
TP248.65.F66 L35 2001
363.19'2—dc21

                                    2001031812

First St. Martin's Griffin Edition: December 2002

10 9 8 7 6 5 4 3 2 1

CONTENTS

Part Three: BACKLASH

Part Four: COMING TO GRIPS

FOREWORD

The pig with human genes seldom rose. Bigger-snouted and hairier than usual, the boar lay in his pen despite the nudging of a normal pig brought in to keep him company. Pig No. 6707 was like few others: He was the prize subject from the U.S. Department of Agriculture's experiments engineering human genes into farm animals.

The year was 1986. Scott Dine, a veteran photographer, and I were witnessing what no journalists had seen before or after: the results of a government experiment to produce "super animals." In buildings made of pink, corrugated metal on the grounds of the government's Agriculture Research Service in Beltsville, Maryland, twenty-two pens held pigs that, as a result of DNA manipulation, carried the human growth hormone gene. The goal: pigs with less fat, which means leaner bacon.

At the time of my visit, research had not succeeded; out of hundreds of tries, human genes had been implanted successfully in only two dozen or so of the test pigs. And only a few, Pig No. 6707 among them, exhibited the higher success of "expressing" the gene in a new generation.

With their bristly hair and wide muzzles, these animals looked nothing like the pigs on the farm owned by my grandfather, H. Ray Fluegel, in Mackinaw, Illinois. In one of the first litters born with the growth hormone genes, a female piglet had no anus or vagina. Some of the experimental pigs were too lethargic to stand, let alone mate. Others developed arthritis and deteriorated.

"The animals are different, you can tell," a government swine researcher, told me. "Later on, some of them develop problems and die."

My visits to Beltsville were part of the research for a series of stories I wrote in 1986 as a Washington correspondent for the *St. Louis Post-Dispatch*. We called it, not very inventively, "The Gene Revolution: Promises and Problems." But the research I wrote about was inventive indeed, and before I filled up a score of pages, I interviewed scientists, entrepreneurs, food experts, and activists to help me glimpse a future when recombinant-DNA sciences, farming, and food might team up to feed the world a bold new diet.

In my travels, I visited Mike Cannell, who would die in a farming accident, and other Wisconsin dairymen who were frightened by a plan on the drawing board for a new genetically engineered hormone that induced cows to give more milk. I began the story: *Mike Cannell makes a habit of touching each of his dairy cows as he walks by them, and they know him. Franny, Flower, Fancy, and Flossy, calves born this year, will expect the same.*

In the hills of West Texas, I tracked down the location of the first outdoor use of a genetically engineered vaccine. About fifteen hundred of Tommy Maddox's pigs had been destroyed after becoming infected with pseudorabies, a herpes virus known as the Mad Itch, courtesy of a fence-busting wild boar. A genetically engineered vaccine preserved the rest of the herd. "The vaccine saved us, I know it did," Maddox told me.

In Middleton, Wisconsin, Winston Brill, then vice president of Agracetus, a start-up biotech company, called the new science "revolutionary." In Minnetonka, Minnesota, Charles Mustant, then president of Molecular Genetics, another new company, waxed giddy over the prospects. The genetic engineering of crops, he said, would become a matter of "profound geopolitical importance."

I spent time with Jeremy Rifkin, the clever and prolific activist who, with his provocative questions about the technology, became a burr beneath the saddle of the gene jockeys. "Certain technologies ought not be developed because their power is so enormous," he told me.

"Gene Engineering Could Remake Farming," read the headline above my first story in the 1986 series. I wrote: *Genetic engineering is expected to bring dramatic changes in the way Americans grow grain, raise livestock and prepare food. These changes will redraw our food delivery system—the biggest industry in the United States.*

I added: *Much promise accompanies genetic engineering. And so do fears—fears not only of scientific unknowns but also of social and economic effects. If these hopes are to become products, biotechnology must survive public scrutiny.*

In the 1980s I wrote about the "deliberate release experiments," as they then were called, when microbes and, later, engineered crops were first transplanted from labs to the soil. I wrote about the formation of the governmental regulatory policy, cobbled together from three agencies without new laws or a central authority.

Then, for over a decade, I all but ignored the technology whose conception I had chronicled. I went back to writing about national politics and, when election cycles ebbed, I investigated environmental abuses, specializing in exploitative dumping around the world. With most of America, I paid little attention when the genetic engineers turned their magic to our daily bread.

I hadn't even known the Agriculture Department had scrapped its failed swine experiments until 1992, when a British Broadcasting Company producer phoned, wanting to talk with me on camera about what I had seen. When the BBC crew arrived at my office with enough equipment to film a movie and spent half a day probing my memory of the misbegotten pigs, I glimpsed Europe's rapt attention to genetic engineering.

By 1998, I decided to see how far genetic engineering had advanced in the twelve years since I had taken its measure. Much had happened beneath our noses, with the American public—and the American news media—all but oblivious. In 1997, just the second year after the United States had approved genetically modified crops, farmers had planted them on more than thirty-two million acres. And in the spring of 1998, the government was forecasting doubled acreage of transgenic crops. The new technology was coming on strong.

By then, also with little fanfare, the government had given companies the go-ahead to sell nearly thirty genetically modified foods, beginning with Calgene's Flavr Savr tomato in 1994. Genetically altered soybeans harvested in the first two years of commercial plantings were showing up in a host of processed foods, from potato chips to cake mix. Experiments by the thousands tested new and more exotic gene combinations on fruits and vegetables from papayas to potatoes.

Meanwhile, in Europe, the jitters had progressed into a case of full-blown fear. Test plots were greeted with protests and sabotage. In making a case for my reporting project, I tagged 1998 as a pivotal year for genetically modified food. Led by Monsanto, life-science companies pleaded with the European Commission for approvals to plant modified corn, among other crops, and accept import of more United States-grown grains. For the world to accept biotechnology, I reckoned it would have to have the European imprimatur.

For more than two years, the *St. Louis Post-Dispatch* granted me license to chase the story. The newspaper's commitment was bold but fitting, for a St. Louis-based company, Monsanto, led the biotech march. The region aspired to become "the Silicon Valley of plant research."

I traveled to thirteen countries on four continents to write about this powerful new technology and its global implications. I visited laboratories and I walked through fields of genetically engineered crops to see for myself the seedlings of change. I prowled European capitals to understand the emerging opposition. I traveled by rickshaw in rural India to gauge acceptance of this latest technology promising to feed the hungry. I traversed Latin American jungles, by bus, foot, and dugout canoe, to see how bioprospectors sought raw materials for the genetic age. I set out to view the technology not as the scientist sees it but through the eyes of people whose diets—and lives—might change.

My timing was good. For as the insurgent crops took root, so did global resistance. What I found, as I suspected back in the 1980s, is a story far bigger than what's happening down on the farm. What I chronicle in these pages is a world-shaping debate about irretrievable changes to the environment; about the relationship between science and society; about what the world eats—and who controls it; about "men playing God," as the Prince of Wales put it; and about global economics. It's a debate so sweeping that it seems certain to dwarf any trade disputes since the advent of the World Trade Organization. What I observed was the formation of a new global politics of food.

The United States, slower to react, finally joined the debate. Americans are accustomed to progress and uninterrupted scientific marvels, and for several years they paid little heed to the clamor in Europe, where anti-biotech sentiments were growing roots like a religion. But Americans began to notice when studies showed potential damage from modified crops to their beloved monarch butterflies. Later, in the StarLink corn scandal, they watched as food fit only for animals landed on the shelves of their groceries through holes in the United States' biotech regulatory scheme.

By the spring of 2001, the American debate had widened to legislatures across the country, with more than forty bills introduced to regulate engineered crops or the labeling of modified food. Still, it remained largely a decentralized debate, bubbling up from the farm fields and grass roots and still easy to miss for those accustomed to national discussions commencing in Washington. In Europe, though, the flames of controversy continued to rage. Literally. In April, arsonists set fire to a Monsanto depot that contained seeds suspected of being genetically engineered. Graffiti on a wall proclaimed: "Monsanto Killer; No GMOs."

Just as there are questions about the safety of modified foods, there are profound hopes. Already this precocious science is reducing the use of insecticides, the most poisonous of farm chemicals. The next wave of modified crops promises more nutritious food—even food that can ward off disease. Apostles of biotechnology promise that their brave new seeds will bring a second Green Revolution, enabling more efficient use of our finite land so as to feed a global population that will increase by one-third in twenty years.

But will society assume the risks? Will consumers have a choice? There are skeptics who believe that we are on the verge of a biological tyranny with life-science corporations driving technological advance the way physicists, propelled by the exigencies of war, changed the world by splitting the atom. The new life-science companies began making permanent changes in agriculture, and therefore the ecosystem, often with minimal public consideration of consequences. This Mendelian magic was arriving in Orwellian fashion.

Can we trust the handful of multinationals buying up the world's seed companies to control food from farm gate to dinner plate? Will we surrender our birthright to dine at Mother Nature's table for a seat at the New Gene Café?

—Washington, D.C., May 2001

Part One

THE NEW GENE CAFÉ

1

ON OPENING DAY, FIELDS OF DREAMS

Will we survive our technologies? We are being propelled into this new century with no plan, no control, no breaks.
—Bill Joy
President, Sun Microsystems

Do you know why people fear DNA? Because criminals always leave it at the scene of a crime.
—Joke told by Monsanto scientist Stephen Rogers

GENETICALLY MODIFIED FOOD is part of the fabric of American life."

So says Gene Grabowski, my seat mate and a front-line player in the new politics of food, as vendors hawk hot dogs, nachos, and Crackerjacks in front of our Section 11 box seats in Camden Yards, one of America's grand new baseball parks. Moments before, the Orioles' Cal Ripken clunked his 2,992d hit as a Major League player into a swath of grass temporarily devoid of any Cleveland Indians in short right field.

"In a grocery, as much as 70 percent of the processed food might contain GMOs," Gene tells me. As a vice president of the Grocery Manufacturers of America and therefore chief spokesman of the American food industry, he ought to know.

GMOs. Grabowski is speaking in a code that most Americans haven't unraveled. In parts of the rest of the world—including Europe, Japan, and Brazil—these three letters trigger fear and befuddlement, with a measure of hope sprinkled in. As most Europeans can tell you, GMO stands for genetically modified organism, which is what you get when you move genes across the traditional species boundaries of plants and animals in the quest for new traits.

It is Opening Day at Camden Yards, and Gene has invited me to watch baseball and, as I suspected, to talk about genetically modified food. The subject has consumed us both of late, he as point man for American food retailers, who worry increasingly about the reaction to GMOs in their food; I as a newspaper reporter writing about a powerful technology that has landed on the world with breathtaking speed. It has been in our midst only since the mid-1990s, the brainchild of a handful of companies that have bigger plans for re-creating what we eat.

Up to now, the DNA of plants has been manipulated to make growing them easier. Companies have profited, and farmers have saved money by heading better equipped into the battle with weeds and insects. But there's been little in the technology to inspire consumers, which is one of the reasons that Gene is feeling anxious today. He would love to see scientists hasten their quest to produce genetically modified food that is more nutritious—or more appealing in any way—so that people won't be suspicious when they learn GMOs have occupied their supermarket shelves.

"So far, we've had to be futurists, talking about the foods that will be available someday, like fruits and vegetables that can retard tooth decay. And that's been one of the difficulties. It's been a challenge, always talking about the future. I like painting a picture of the future, but it's always easier when you have something that is concrete," he tells me, as we alternate between baseball and GMOs during this annual rite of spring.

I joke that in my mind, it's not *really* Opening Day, seeing as how Major League Baseball commenced its season in Japan five days earlier. Hoping to enhance the game's global appeal, baseball marketers dispatched the Chicago Cubs and the New York Mets to perform the Opening Day ritual on foreign soil. To dedicated fans, this was heresy. But tinkering with baseball is inconsequential compared to the bold drive by corporate science to reorder the world's food system. At the moment, they are succeeding, albeit neither as swiftly nor as stealthily as they had hoped.

ex's of GMO's across the world

Fans watching Major League Baseball open its 2000 season at the Tokyo Dome ate snacks that contained GMOs. If they dipped their sushi, they undoubtedly consumed soy sauce from genetically modified soybeans grown in the United States. In China, hundreds of thousands of cotton farmers had sown modified seeds the season before, and the government also had commercialized engineered tomatoes, cucumbers, and a pepper variety, in addition to its engineered tobacco. In Argentina, the vast majority of seventeen million acres of soybeans were genetically engineered. In 1999, three new countries—Portugal, Rumania, and Ukraine—planted engineered crops commercially for the first time, bringing to an even dozen the countries of the world where they legally sprout. Even Europeans, who by and large spurn the technology, were, whether they like it or not, eating food processed with genetically engineered soybeans.

When it comes to transformation of food, Americans lead by example. Ball Park Franks, a brand of hot dogs, was one of many foods found to contain genetically modified ingredients in tests sponsored by *Consumer Reports,* the magazine, and advocacy groups. As Gene had suggested, genetic engineering is as American as the national pastime.

North Americans are eating genetically modified foods regularly, but they don't know which ones because, unlike Europe, Japan, and Australia, the governments of the United States and Canada don't require labeling that provides this information on food packaging. Thus, North Americans are unaware of how deeply the technology has already reached into their cupboards. Tests by the consumer groups also showed altered DNA in breakfast cereals; corn and tortilla chips; granola bars; cake and muffin mix; corn meal; diet drinks; dog food; soy burgers; powdered chocolate drink; and taco shells. The new modified diet starts young; GMOs were found in three types of baby food.

GMOs are drunk as well as eaten. At Camden Yards, Gene reminds me that cola and soft drinks contain high-fructose syrup made from bulk corn that is likely to have engineered hybrids mixed in. Dairy farmers are using a genetically engineered hormone that induces cows to give more milk. Modified milk blends in the general supply of the beverage that's

CON for public

hired wholesome hero Cal Ripken as its poster boy. Next, barley breeders intend to use genetically engineered varieties in beer. Scanning the patchwork of reds, yellows, and Oriole orange worn by fans in the rows in front of us, Gene observes that many in this crowd of 46,902 are wearing cotton from genetically engineered plants.

Our genetically engineered food is new, so new that on September 6, 1995, the day that Ripken surpassed Lou Gehrig's "Iron Man" record of 2,130 consecutive games, gene-altered corn and soybeans had not yet been planted commercially. They were sprouting in American fields for the first time the following spring, when Ripken broke Japanese third baseman Sachio Kinugasa's world record of 2,216 games.

GENETIC ERA DAWNS

On October 19, 1992, the U.S. Department of Agriculture approved Petition No. 92-196-01P, which allowed Calgene Incorporated to proceed with commercializing its Flavr Savr Tomato. Two years later, Flavr Savr became the first genetically engineered product to reach U.S. supermarkets. By then, China was already producing tomatoes and tobacco, after having sown its first commercial crop in 1992, a transgenic tobacco resistant to the cucumber mosaic virus, on approximately one hundred acres. Two years later, Chinese scientists had engineered a second gene into tobacco to ward off tobacco mosaic virus.

By 1996, the U.S. Agriculture Department had approved more genetic variations of Calgene's invention, in which a gene was inserted backwards to slow the speed at which tomatoes softened as they ripened. The manipulation was supposed to remedy the tastelessness of tomatoes picked long before they're sold. Unfortunately for the genetic engineers, the Flavr Savr tomato was as short on consumer appeal as on vowels. What we grow in our gardens remains the standard for comparison, and even gene wizards couldn't produce a tomato that good.

The first truly revolutionary crop genetically engineered in the United States, a Monsanto Company soybean, won the government's blessing on May 19, 1994, ushering in a series of government approvals for corn, potatoes, more tomatoes, cotton, squash, papaya, and, oddly, radicchio. In 1996, the first year GMO crops were grown commercially, American farmers planted 3.6 million acres, surpassing China. In Canada that year, farmers planted about 300,000 acres with an herbicide-tolerant canola. Argentina, Mexico, and Australia had also begun cultivating a small acreage of modified plants. But nowhere would the new crops proliferate as in the United States.

By the time the new century arrived, the American government had approved more than fifty bioengineered crops. In 2000 in the United States, soybeans, corn, potatoes, and cotton were cultivated on seventy-five million acres of the 109.2 million planted globally. Never before had the worldwide acreage exceeded one hundred million, a landmass twice the size of the United Kingdom. The vast majority of these crops had genes inserted for two traits: herbicide tolerance, which enables plants to withstand sprayings of proprietary herbicides, primarily Monsanto-created Roundup formulations; and insect resistance, which equips plants with the gene of a bacterium, *Bacillus thuringiensis*, so that they produce a protein that is fatal to pests.

Already, thousands of processed foods around the world contain genet-ically modified ingredients, most often modified soybeans. In the vision of the life-science companies, that is just the beginning. The seed catalogue of modified foods tested in the United States is thick indeed. In thousands of experiments during the century's waning years, companies and univer-sity scientists conducted tests engineering new traits into wheat, rice, can-ola, melons, squash, cucumbers, strawberries, and sugarcane. Into apples, coffee, cranberries, eggplant, oats, onions, peas, pineapples, plums, rasp-berries, sweet potatoes, walnuts, and watermelons. Science is marching us toward a new gene smorgasbord, with many foods seasoned with DNA that has never before existed in the supply of human food.

It doesn't stop with food. Modified tobacco has been tested outdoors, and experiments have been conducted manipulating the DNA of creeping bentgrass, Kentucky bluegrass, the American chestnut, spruce trees, sweet-gums, geraniums, gladiola, and the Texas gourd. Once approved by the Agriculture Department, these outdoor tests proliferated. In 1987, there were just five of these field-test sites approved. Then:

1988—16	1995—3,859
1989—40	1996—2,997
1990—81	1997—3,792
1991—155	1998—5,088
1992—381	1999—5,102
1993—905	2000—4,549
1994—1,926	

ex's of potatoes + = monsanto

No longer are companies content to add a single gene. Monsanto, which remained the leader in the gene-altering race, has added as many as eight to potatoes. These "stacked-gene" potatoes provide resistance to pests and diseases, add tolerance for direct applications of herbicides, increase solid content, and reduce bruising.

PRO for farmer

Smelling french fries at Camden Yards, I recalled to Gene Grabowski the words of an Agriculture Department biotech expert, Arnold Foudin, when we talked about these experiments: "That's a potato that can take care of itself."

Summarized

"OPENING DAY" ALL AROUND

The genetically engineered NewLeaf Potato, as Monsanto called it, may be so tough that it can kill bugs, so resilient that it can ward off fungi, so ruggedly constituted that it can withstand the vicissitudes of transport, so muscle-bound that it weighs in for market heavier, therefore netting the grower more. But so far, there's been no gene discovered to help the potato handle its public relations.

That would become apparent in the spring of 2001, when Monsanto acknowledged that it was bowing out of the business of genetically modified potatoes. With Monsanto holding iron-fisted control over the gene-altered spud technology, the company's concession to the emerging new politics of food meant that no modified potatoes would sprout in the United States and Canada in the next growing season. "We hope to return to it some day. For now, the potatoes will be mothballed," a company spokesman told me, prompting me to ponder for an instant how a baked, genetically engineered potato that had been stored in mothballs would taste.

With so many foods modified so soon, the creators of genetically modified food have led us to believe that the march of biotechnology is unstoppable. But the future is much less certain than Gene Grabowski's grocery statistics might suggest. A backlash against GMOs in Europe has spread to other continents and cultures and sprouted in the United States. The reaction was rooted in worries about safety; about the control of food in the hands of few companies; about a new technology with the power to reorder the building blocks of life.

This April day, farmers also were opening their new seasons, heading into the fields. The roar of rejection from Europe, accompanied by new chords of disapproval elsewhere, left them wondering and worrying. Would they find buyers abroad for their harvests? How, when the breadth of food-changing became widely known, would American consumers respond?

By spring 2000, the acreage of American soybeans sown in genetically engineered seed had increased to about 54 percent, while modified cotton had claimed 61 percent of cotton fields. But a 20 plus percent drop in engineered corn testified to spreading fears. For the first spring since 1996, when genetically engineered crops had become legal, sales of the new crop wonder had fallen.

For the global biotechnology industry as well, this day, April 3, 2000, was about more than baseball. This was Opening Day for their new offensive to hold back the tide of opposition. On this morning, seven life-science companies—Monsanto, Novartis, DuPont, Dow Chemical, Zeneca Ag Products, Aventis CropScience, and BASF—announced that they had formed an unprecedented alliance, committing $50 million for a yearlong information campaign in North America. By 2005, their spending in defense of GMOs may reach $250 million, testimony to the enormity of the coming battle.

This April day, they opened a coast-to-coast television campaign heralding the rewards of their new technology. Before heading to the ballpark,

I previewed the first spot, which recalled to me the "Morning in America" feel-good commercials in the reelection campaign of former President Ronald Reagan that I had covered in 1984. A boy and his dog, a golden retriever, loping together, fade to a farm girl with two calves. A voice alternates between telling us of successes down on the farm and trumpeting breakthroughs in medical research. "Discoveries in biotechnology, from medicine to agriculture, are helping doctors and farmers to treat our sick and to protect our crops," we are told.

In thirty seconds, I identified Caucasians, Africans, and Asians; farmers, scientists, doctors, and athletes; dogs, cattle, seagulls, and geese. The message was, indeed, Reaganesque: Biotechnology is bringing a new day to America. Amid music that soothes, we're told that genetic engineering of food is no different than the techniques that make our medicine.

Sponsors of these ads have invested billions of dollars to create the recombinant-DNA technologies that farmers carried into their fields on this day. They had purchased the seed companies that sell farmers what they plant. They had budding monopolies along the food chain—or so they thought. Suddenly, the backlash had rendered those investments risky. No company was feeling the pressure more than Monsanto, the band leader of the biotech march, for whom this, also, was a new day.

For Monsanto, the pioneer of the bold new technology, the company that looked to all the world to be toting the shotgun at the marriage of genetic engineering and agriculture, April 3, 2000, was unlike any day in ninety-nine years. Since 1901, Monsanto had stood alone, prospering near the banks of the Mississippi River, first as a chemical company, then reengineering itself into a hybridized life-science company. But this was the first business day after a merger that has diminished its stature; now Monsanto was a subsidiary of Pharmacia Corporation.

Gene Grabowski's hope is that American consumers won't demand that genetically engineered food be labeled. Block labeling. Squelch the opposition. These are the imperatives for the biotech and the food industries, which are allied in battle. That is why, in talking about what people eat, I am hearing words that describe how people fight.

WAR OVER WHAT WE EAT

Grabowski matches wits with consumer and environmental activists from his office overlooking the Potomac River at Georgetown. From cake mix to Spam, the Grocery Manufacturers of America keep the goods of its members displayed behind glass like artifacts at the Smithsonian. With 142 affiliates—from giants like Kraft, Kellogg's, and General Mills to pint-sized operators like McKee Foods, of Tennessee, maker of L'il Debbie snack cakes—the association is the world's biggest trade group for food. The companies Grabowski speaks for sell $460 *billion* worth of products each year in the United States alone.

Gene brings to the job a pugnacious attitude that must run in the family. He is a cousin of football great Jim Grabowski, whose career as a high-stepping running back took him from the University of Illinois to the Green Bay Packers. Speaking about his industry's campaign, Grabowski sounds like a field general, or perhaps like the Packers' legendary coach, Vince Lombardi. "They hit us with everything they had, and they couldn't put us down," he says, describing the efforts of opponents to genetically modified food. "Now, we strike back."

This is not a war the food industry started. Monsanto and its rivals did not seek permission of food retailers of the world before engineering DNA into foods, patenting the outcome, and fanning out into farm country offering promises to farmers to sign contracts to use the new technology. But it's a war the American food industry must fight, just as Europe's food industry fought it and, for the most part, lost, in the waning years of the twentieth century. At this moment, Grabowski and the American food manufacturers viewed the debate over mandatory labeling as the primary battleground. So far they had won, persuading the government and its Food and Drug Administration to resist entreaties to let consumers know the derivation of what they are eating.

But the fight is bigger than tiny words on the side of a package. At stake in the coming years is the freedom of companies to move genetically engineered foods around the world absent restrictions never before applied in the commerce of commodity foods. At stake is the business structure of agriculture, with farmers required, for the first time, to pay technology fees and sign contracts to plant the brave new seeds. At stake is both the content and the structure of our food system.

It was a war easy still to miss in the spring of 2000. But a flurry of developments during the new millennium's first months gave evidence that a new politics of food had migrated to the United States from foreign soils. McDonald's, the American-based paragon of fast food, ordered suppliers to cease delivering genetically engineered potatoes, those invincible spuds for all seasons. Fearing consumer rejection, McDonald's was following the example of European companies. Another American company, Frito-Lay, had made a similar announcement about potatoes, three months after advising its corn suppliers against delivering bioengineered grain. Both Heinz and Gerber had announced their intentions to remove GMOs from baby food. Novartis, the Swiss company that owns Gerber, was on the verge of directing the removal of modified ingredients from all of its food product lines. In other words, one of the life-science behemoths directing the biotech revolution had, itself, backed away from the end result of the technology.

Meanwhile, alliances of shareholders were pressuring two of Gene Grabowski's stalwart members—Kellogg's and Safeway—to refuse modified products. Even more spirited protest sparked in Washington when anti-GMO campaigners joined in the mass demonstrations to shut down the annual meeting of the World Bank and the International Monetary Fund. The protesters failed to stop the meetings, but they retreated to fight again over the basics of life.

Sabotage, sprouting in the United States as it had in Europe, was the boldest shot of all. That troublesome reality had displayed itself when the Genetic Century commenced with more than the usual fireworks.

SABOTAGE

On December 31, 1999, Catherine Ives made a copy of every computerized file she would regret not seeing again. With worries rampant about Y2K computer failures, she wanted to protect the records of genetic-engineering research at Michigan State University's Institute for International Agriculture. She put diskettes with all of her duplicated files in the drawer of the desk in her office. Then she left, planning not to return to Agriculture Hall until the new millennium.

Ives managed her university's Agriculture Biotechnology Support Project, which is funded by the United States Agency for International Development to help developing countries secure genetic sciences for their labs and modified plants for their fields. In research at Michigan State, scientists have reported successes manipulating the genes of potatoes, melons, and squash with the aim of sowing these transgenic crops in countries that want them.

In Egypt, researchers from Ives's university were helping in the fourth year of field trials testing potatoes engineered with a gene for *Bacillus thuringiensis,* otherwise known as Bt, to ward off the potato tuber moth. In another year or two, they hoped to see insect-resistant potatoes planted commercially. Michigan State has worked with a similar gene engineered into corn in Indonesia, as well as with fungus-resistant sweet potatoes in Kenya.

Since 1991, Michigan State had spent about $20 million, most of it from the federal government, to hasten the arrival of the genetic era to the world's least-developed countries. The university works not just with the science of biotechnology but with its bureaucracy, helping countries set up the regulatory machinery to oversee a new brand of agriculture. No other American university has been so energetically involved.

On New Year's Eve, feeling secure that her records were safe from computer glitches, Catherine Ives headed out of her office at Michigan State University to party. Shortly before eight o'clock, she was walking along Grand River Avenue on her way to a local watering hole when she saw fire engines. *Looks serious,* she thought. As she walked, it seemed like the sirens' screams were ending near her office. It couldn't have been nearer.

She quickened her pace, detouring several hundred yards onto the campus. When she arrived at Agriculture Hall, flames were leaping from the fourth floor, out of a window from which she had peered many times: the window of her office. It was a destructive fire in one of the oldest buildings on campus. Damage was estimated at $400,000 at the time, but Ives believed it to be at least twice that. The labs were not hit; the program assisting developing countries was delayed a few weeks, but the genetic-engineering research continued. As far as records, it was a different story. The losses to Catherine Ives's files, notes, and just about everything connected to her profession were total. Those precious diskettes melted in their drawer.

"There was nothing left to my office or my associate director's office," she recalled. "Everything I'd collected over ten years, you name it, it was gone. For an academician to lose their research materials, it's just devastating."

Unaware of the fire's cause, Ives hurried home to telephone people she worked with. Agriculture Hall was old, built in 1909. But it had just been spruced up with an eight-million-dollar renovation. She thought it might have been bad wiring. She didn't consider the possibility of arson, even though Michigan State had been a target before: In 1992, animal rights activists lit a fire at nearby Anthony Hall and set lab animals free.

Three weeks later, investigators had more than a clue. The Earth Liberation Front, a loose-knit activist organization that has orchestrated attacks for several reasons, took credit for the New Year's Eve blaze. There was speculation that Michigan State had been targeted because Monsanto had chipped in two thousand dollars to help pay for travel of some students to a biotech conference. In the minds of activists of many stripes, Monsanto is "Monsatan," as proclaimed on one of the radicals' web sites.

Whatever triggered the attack, Americans had a signal that a powerful technology might not be arriving smoothly in the new millennium. And Catherine Ives had her eyes opened to a reality she could not have conjured. "To me, it just showed a tremendous amount of ignorance. We're not even funded by Monsanto. From now on, I certainly will be much more skeptical of any environmental organization, and I will question their motives. And that's a shame, because I consider myself a conservationist," she said.

Catherine Ives never made it to the bar. She did her New Year's Eve drinking at home.

BREAKTHROUGH

The fortunes of war swing both ways. In the spring of 2000, the industry enjoyed a public-relations bonanza from a wave of reports about the successful engineering of beta-carotene enriched rice, an achievement that could help the world's million children weakened by vitamin A deficiency. And even as Monsanto was relinquishing its independence, the company made a stunning announcement: Its researchers had completed the first working draft of the genome of rice. Together, these discoveries carried the potential for improving the nutritional value and yield of the world's major food crop, thus fulfilling biotechnology's brightest promise.

The genetic engineers and the food industry had a smaller but significant victory to celebrate in that pivotal month of April when the National Academy of Sciences concluded that the food from crops modified for insect resistance is safe. It marked the first of a series of benchmark endorsements from global scientists that would boost the morale of Grabowski's grocers and gird the industry for battle.

In my conversations with Gene Grabowski, I have learned some of how the world's biggest industry—food—operates. I could be wrong, but I thought that Gene, a former journalist, sensed an opportunity to learn from me about the politics of food I'd studied in my travels to the front lines of the global debate. Seldom in my dealings with the biotechnology industry have I seen company officials willing to reach beyond their coterie of advisors, lawyers, and hired-gun public-relations specialists to understand what is happening at the grass-roots level. I think that this ostrich attitude may have been a costly mistake, one that may or may not be repeated now that the debate has reached American soil.

Companies bringing these genetic technologies tell us that modified foods are simply a natural progression of a science, classical breeding, begun when a shy, portly Austrian monk, Gregor Mendel, all the while smoking cigars and feverishly writing notes, crossed round peas with wrinkled peas, tall plants with dwarfs.

The biotech industry will be spending millions of dollars in the next several years to persuade us that food biotechnology is a life-sweetening advance in the Mendelian tradition to stand alongside the deciphering of the human genome and the creation of new medicines as triumphs of the Genetic Age. In sanguine moments, the corporate scientists say that decades from now, the genetic revolution in our food will be ranked in the same class as the computer revolution among the twentieth century's most profound inventions. They may be right.

But inventions don't always succeed, as we saw in the 1970s when Congress declined to spend money on a supersonic transport. Pulitzer Prize–winning scholar Jared Diamond has drawn a list of reasons why some inventions succeed and others do not. On top of that list is the relative economic advantage compared with existing technology. Given what farmers are reporting about lowering costs for pest control, biotech would pass. Another factor is the invention's compatibility with vested interests. As far as melding with the system of commodity farming, biotech seems to pass that test, too, although it remains to be seen whether the food industry ultimately will buy into the revolution.

Another consideration is clarity of benefits. Biotech promises healthier food and a cleaner environment, advantages that one day may be clear to the public. So far, they're not. From my travels, I know of no threat to human health from the act of genetic engineering, whether it be performed on a potato or a papaya. But I have heard many questions. Is the new food safe for humans? Will introducing these plants damage the environment? What will be the outcome of widespread crossing between genetically engineered crops and their wild relatives? What weapon will organic farmers and gardeners have if pests develop resistance to Bt as a result of its overuse by the genetic engineers? What about the fate of such unintended targets as the queenly monarch butterfly when plants are engineered to kill insects?

Still another criterion is social value, and on this score, society is still making up its mind. In the realm of social questions, how will we assure that the products of biotechnology and its patented techniques are distributed fairly in the world? Are there broader implications for society in allowing a handful of companies to reorder the building blocks of much of our food supply? And, individually, will we have a choice to refuse genetically engineered foods—without paying higher prices at whole-foods markets?

There are equally compelling questions on the other side of the ledger. Does biotechnology truly have the capacity to help feed a global population that is expected to swell from six billion to eight billion by 2030? What about the demands by people rising out of poverty for more meat and more oil in their diets? Will genetic engineering allow us to grow more food on the world's six million acres of arable land, so that we don't have to destroy even more of the rain forests that keep the global ecosystem in balance? And will it enable us to make real progress in weaning the world away from farm chemicals?

These are the questions that will be reverberating in the United States in the coming years as people make up their minds about eating the new diet the biotechnology industry wants to feed us. Many of the questions in this debate are profound. Will the ringmasters of biotechnology succeed in their drive to modify the world's food supply? Is this epochal change inevitable, as we are being told? As Americans decide, the chasm will deepen between the products of an increasingly intensive agriculture and the organic wares from more natural methods of growing.

Gene Grabowski knows this, and he admits that the industry's strategy is a stalling game: Hold off labeling until genetic engineering brings healthier foods and modified products that appeal to consumers, not just farmers. And then, as the industry's thinking goes, today's growing opposition will melt into the margins.

"The longer time we have before a rush to judgment, the better off we are," Gene tells me.

My own belief is that GMOs are here to stay, barring unforeseen health threats. I also think we're headed toward a two-track system: There will be GMO food and non-GMO food, and the non-GMO food system will need determined, fail-safe preservation from farm field to supermarket with separate farm implements, grain elevators, processing plants, and even ships, in much the same way kosher food is kept untainted. I tell Grabowski that I think it unlikely that the industry will be able to hold off mandatory labeling of its modified products much longer. Americans entered this millennium able to eat what they want and newly empowered by the same political currents that propelled third-party politics in the 1990s. People want a choice, and they will get it.

He disagrees, and in his candor he lays waste to the inevitability argument while hinting that food sellers and the biotech industry might not always view the world in the same way. In his view, we will have a general system of modified foods, or we won't. It's that simple, and it comes down to economics.

"It's economy of scale. You can't have a two-track system. It's like Beta and VHS. Beta is the superior technology, but VHS won the battle. Choices have to be made in the market. Nobody can afford to efficiently and affordably provide two consumer products. If you have to segregate out the products to make sure they are not biotech, it would just be too expensive. Farmers and food manufacturers are just going to have to look at this situation and decide whether it's going to be Beta or VHS. We'll either go biotech, or we won't," he says.

"This is a war that we will win—or that we will lose."

A war over food.

PLANTINGS ONE ⬤

Mud streaks my wrists like a double helix and cakes my bare feet. Just before the rain arrived, I completed a dirty task and one that could be portentous for humankind, I am told. There is no question that what I've done is illegal.

Moments ago, I planted genetically engineered soybeans— pirated genetically engineered seeds, to be exact—in my own backyard. I didn't deploy a drill unit or a gauge wheel or any fancy planting device, like a real farmer. I crawled in the soil with a rubber-handled trowel in my right hand and as many of the beans as my left fist would clutch. I may have spilled some in my tomato patch.

I kept these illegal beans in my truck for many months, and I wondered whether they would still be good. Then I read about an Englishman named Ramsbottom who, in 1942, germinated Nelumbo seeds from a sample in the British Museum that had been collected in 1705. A Japanese scientist claims to have sprouted thousand-year-old seeds found in oxygen-deprived peat in southeastern Manchuria.

By comparison, my soybeans are veritable upstarts: the technology that endowed them was not legal in American fields until 1996. So I do not worry about the viability of my fertilized little ovules. My gene-altered beans looked like black-eyed peas: beige, egg-shaped BBs with black eyes casting glances about. The high-tech beans didn't jump out of my hand, like those Mexican beans. They didn't burrow into my sweaty skin or burn me. They behaved, at least until I covered them with a couple inches of Maryland soil.

Alongside them I planted a row of Grandpa Ott's Morning Glory, a robust Bavarian seed that grew fifteen feet high on the south side of the barn at Heritage Farm, in Decorah, Iowa, where they came from. Behind them I put in a patch of Moon and Stars watermelon seeds, which yield giant green fruits with spooky markings. I figured that together, Grandpa Ott and the cosmic melons can control my genetically engineered soybeans. Just to be safe, I planted them where I can keep an eye on them.

But I worry that someone might be keeping an eye on me after I re-read an old news release from Monsanto Company, creator of gene-altered beans, warning that the company might deploy satellites to track its technology. I'm not sure if their satellites are pointed toward where I live, along the Chesapeake Bay, or if they can penetrate the fortress of maple trees surrounding my home. I do know that in the past, Monsanto has hired detectives to track pirates, and I suspect that they could find me.

My case would not be strong. I paid no one a "technology fee," nor did I sign a contract agreeing to use the seeds for one crop only and never, ever to save them. I will not say from where they came because he (or she) could get in trouble, too.

The modified beans I planted are engineered for herbicide tolerance. The label on my ill-gotten beans says: Roundup Ready Gene. That means that a gene was spliced into an earlier generation enabling me to dump weed killer over them abundantly, and they will not die. But not just any weed killer; only Monsanto-created glyphosate, known to the world as Roundup.

Why did I plant them? Around the world, a debate has exploded about genetic engineering in farming and the wisdom of eating modified food. At present, where the tractor's tires meet the dirt, what this technology is mostly about is herbicide tolerance. Of all the genetically engineered seeds planted in the world during the first season of the new millennium, more than 70 percent were put there to manipulate the workings of weed killers. Nor is this technology yet about "golden rice" or vaccines from bananas; soybeans make up nearly 60 percent of the global plantings of modified crops.

So I will conduct my own field trials to see what this technology is about. Alongside the gene-altered soybeans I have planted conventional beans. We will have a competition.

2

IN THE BEGINNING . . .

In the beginning God created the heaven and the earth. And God said, Let the Earth bring forth grass, the herb yielding seed, and the fruit tree yielding fruit after his kind.

—Genesis, 1:1 & 11.

. . . thus render ourselves lords and possessors of nature.

—Descartes

MOLECULAR BIOLOGIST ROBERT T. FRALEY let the single sheet of paper float from his fingers onto a desk at Monsanto's headquarters at the junction of Olive Street Road and Lindbergh Boulevard just outside St. Louis proper. His special-team colleagues looked at the page he had delivered to them. They looked again. They were dumbfounded.

"Expression of Foreign Genes in Plant Cells," read the headline on a cover sheet bearing the identifier of the National Academy of Sciences. When the words sank in, the Monsanto scientists understood that, for the first time in history, somebody had genetically engineered a plant.

"We've been scooped," teammate Robert Horsch blurted out.

It was 1982, and Monsanto had waged its own drive to achieve this scientific breakthrough. Hidebound captains of the staid old chemical company had been persuaded to mobilize a cadre of highly specialized scientists for a project that was fascinating—though its benefits weren't altogether clear. In "U" building at the company's Creve Coeur, Missouri, laboratories, they worked nearly every night, hungry young men on a mission. Nearly a decade before, scientists in California had successfully transferred genetic material between organisms. It was one of the series of discoveries that followed the groundbreaking work of James Watson and Francis Crick who, in 1953, had deciphered the double helix structure of DNA while working at the University of Cambridge. But no one had engineered a plant, any plant, so that it expressed the desired characteristic and then bequeathed its new traits to the next generation.

Until now, apparently.

The crestfallen Horsch weighed in his hand the sheet of paper his colleague presented. "Who *are* these people?" he asked, indignation rising in his voice. Horsch was not even thirty, but he knew the heavyweight plant scientists, at least by reputation. He recognized none of these names. He read more, and the words began to look . . . fishy. His eyes shifted back to the top of the page. Part of the title seemed fuzzy, askew. Horsch and the others are scientists, not detectives or graphologists. But in a few moments they realized that they had been had. Fraley had *spliced* words on the page to fool the splicers.

The Monsanto researchers went back to work.

A BRIEF HISTORY OF A COMPANY IN THE CROSSHAIRS

Monsanto is, unquestionably, the world leader in the genetic engineering of food. It's been *the* company with the foresight to see the horizons of biology and the steadfastness to try to reach them. It was also Monsanto that pioneered the business innovations of food biotechnology: the life-science structure of companies; licensing agreements with rivals for its seed technologies; user agreements with farmers requiring them to pay technology fees; and the aggressive use of patents to protect their discovery as intellectual property. For all its efforts, Monsanto has endured an extraordinary barrage of skepticism, criticism, and outright condemnation unleashed around the world on the arrival of genetically modified food.

In the waning years of the 1990s, Monsanto mutated in the minds of many from a typical American company, albeit an eight-billion-dollar one, to a corporate demon. Indeed, "Monsatan," was coined as a suitable *nom de guerre* for a company with the temerity to cross the boundaries of species to set the global table and upset the food business. From the outset, Monsanto's detractors seemed unmoved by biotechnology's potential benefits to the earth of using fewer chemicals and saving soil. Often, they ignored other multinationals that had charted a similar path. Monsanto—"Mutanto," another choice label in the era of the Internet—was the lightning rod.

The firestorm erupted a century after Chicago-born John F. Queeny settled in St. Louis in 1897 to begin a new job as purchasing agent for a drug firm. Queeny had not advanced as far as high school, but he had trained himself as a chemist and he harbored the dream of owning a company trading in chemicals. That dream was fulfilled two years later, albeit fleetingly. His new sulfur-refining plant, in East St. Louis, on the Illinois side of the Mississippi River, burst into flames on the day it opened. His life savings went with it.

The indomitable Queeny, known for his ever-present cigar, wouldn't give up. He persuaded investors that they could profit from the manufacture of an artificial sweetener that had been imported from Germany. In downtown St. Louis two years after the fire, he christened the Monsanto Chemical Works, borrowing his wife's family name.

Queeny's fledgling operation survived efforts by German chemical companies and the American sugar industry to chase it out of the sweetener business. Monsanto expanded first into caffeine and later, amid the prosperity of war, into sulfuric acid and a variety of other chemicals. Tongue cancer, possibly from the cigars, killed John Queeny in 1933. But his son, Edgar Monsanto Queeny, stepped in with an aggressive vision that expanded the company east to Massachusetts and south to Alabama.

World War II delivered a new measure of prosperity for the company, which had begun producing styrene monomer, a chemical component of synthetic rubber. When the war ended, Monsanto was a hundred-million-dollar company looking for new directions. The firm turned to retail products. In the 1950s, Monsanto sold low-suds detergent and soil conditioner and even tried its hand in the petroleum business by purchasing Lion Oil, an Arkansas company. One of its most famous products, Astro-Turf, arrived at the Astrodome in Houston in 1965. In that decade, Monsanto became a one-billion-dollar business and streamlined its name to Monsanto Company "We've outgrown our middle name," ads said.

By now, Monsanto had begun to concoct the killing compounds that would one day tarnish its name and begin to fill a reservoir of public suspicion. In 1935, Monsanto had bought the Swann Chemical Company, of Anniston, Alabama, maker of polychlorinated biphenyls, better known as PCBs, which were used in electrical transformers and capacitors as well as in pesticides, flame retardants, and lubricating oils. Monsanto also produced PCBs at its W.G. Krummrich Plant in Sauget, Illinois, across the river from St. Louis, for about thirty years. From 1935 to 1977, Monsanto had the distinction of being the sole American manufacturer of PCBs, which were banned after studies on laboratory animals demonstrated their carcinogenic properties.

In the Vietnam War years, Monsanto was among the companies that produced Agent Orange, the herbicide containing 2,4,5-T and 2,4-D, which was sprayed in Vietnam to destroy foliage and therefore make it difficult for the enemy to hide. In 1984, Monsanto and six other companies agreed to a $180 million settlement in a court case brought by thousands of veterans who blamed Agent Orange for cancers, birth defects, and a host of illnesses.

In my early dealings with Monsanto, as a reporter in Illinois, I disclosed in 1982 that a seeping Monsanto chemical dump across the Mississippi River from St. Louis was among the sources threatening local waterways with a gumbo of toxics. The Illinois attorney general later sued the company to hasten cleanup, remarking at that time that the contents of the dump "read like a shopping list of toxic and hazardous chemicals . . . phenols, lead, chromium, dioxins, substances that are highly toxic to human health or animal life."

I tried to balance the picture by writing that the nearest intake pipe for drinking water was far downstream. I pointed out, too, that Monsanto had not been guilty of criminal acts; the dumping had occurred over three decades before society rose up to declare that toxic pollution no longer would be acceptable. Nonetheless, I wondered why a company that claimed to have a stable of the world's finest scientists, people knowledgeable about the threats of their wares, would dump dangerous chemicals in ditches just because they could.

The litigation dragged on for years, and the list of contributors to the pollution grew to more than two dozen. Finally, in the waning days of 2000, Solutia, a chemical company that had been spun off by Monsanto, voluntarily commenced excavation and cleanup of a swath of earth that had come to be known locally as "Dead Creek."

Monsanto would prefer to be remembered for its chemicals that helped farmers grow food, especially its world-famous Roundup herbicide. But as far back as the 1960s, a few people in the company already were wondering if it would take more than chemicals to keep Monsanto flying high.

ERNIE JAWORSKI

Fraley's phony-document gag was testimony to the cooperative spirit of the Monsanto team racing to become the first scientists on the planet to genetically engineer a plant in a systematically reproducible way. In that pathfinding endeavor, cooperation was a key: The mysteries that had kept species on this earth separate from one another were profound, and crossing the boundaries of natural systems took cutting-edge approaches in multiple scientific pursuits—all at one time.

Crossing those boundaries also took a leader who could keep his eyes on the prize while marshaling his talented team and running interference with the corporate bosses. He was Ernest G. Jaworski, a biochemist in Monsanto's agriculture division and self-described "player-manager" for the team.

Jaworski carried the reputation in his company as a maverick for his willingness to stray outside the corporate corral of thinking. In the late 1960s and 1970s, he had begun to peer into the future to a time when "you've invented all the herbicides you need, all the insecticides you need, all the fungicides." What then would the company do for its daily bread?

"I call it a constructive dissatisfaction with the status quo. You need to be ready for change, because change is a constant," Jaworski, who retired from Monsanto in 1991, told me when we met in St. Louis at Coco's Restaurant, down Lindbergh Boulevard from Monsanto's world headquarters.

Back in the sixties, "constructive dissatisfaction" was not the accepted motivation in corporations in a city known for its "scrubby Dutch" conservatism. The company's herbicide alachlor, better known as Lasso, had scored rapid success after hitting the market in 1969. Seven years later, another herbicide, Roundup, raced past alachlor on its way to becoming the biggest selling weed killer and a success story nonpareil in the history of farm chemicals.

As he looked into the future, Ernie Jaworski was not disloyal, not in the least. He was simply voicing what his gut told him: that Monsanto needed to think about a new direction, about doing more than one thing, its chemical thing. Perhaps it was in his blood, being the son of a Polish immigrant who was known in his neighborhood of south Minneapolis as "Mr. Fix-It." Leon Jaworski lived up to the name, putting the hum back in anything from one neighbor's Ford to another's refrigerator.

Ernie was born in January 1926, nine months after his father and mother, Mieczeslawa, arrived from Lwow, Poland. Leon Jaworski was fascinated by how things worked, particularly radios and the newest technological marvel, television. He would take young Ernie to the Minnesota State Fair, where they'd track down the scientific and mechanical exhibits. But Minnesota's fair also was a grand agricultural showcase, and all those farm marvels must have stuck with the young man through his service in the Navy and his schooling at the University of Minnesota and Oregon State. Ernie Jaworski's generation, more than some that followed, believed in the capacity of science to relieve humankind's misery. It was that abiding faith in progress that helped Jaworski in 1952, when both his wife and daughter contracted polio.

At Monsanto in the 1970s, chemistry, not biology, remained the ticket to prosperity. To check out their suspicion that spraying chemicals on plants could trigger growth and turn on other traits, the company had devoted extensive basic research to the study of plant growth regulators. But after more than a decade of expensive research, the project was scrapped. Scientists realized that regulating plants was more complicated than they thought. They didn't know which of the plant's genetic systems they were affecting, or how. They concluded eventually that unless they could control climate and sunlight, which even mighty Monsanto hadn't mastered, dictating the internal physiology of a plant would remain out of their grasp.

Jaworski had been concentrating on cellular structure and tissue culture—growing organisms in special mediums. In 1972, he took a two-month sabbatical to study advanced techniques of tissue culture under Olaf Gamborg at the University of Saskatchewan. Not long after he returned to St. Louis, he was planning how to engineer plants. There was no clear idea back then how a chemical company would make a business of biology. But Jaworski understood as well as anyone at the company how Roundup kills plants, by blocking the pathways along which it creates life-sustaining amino acids. He had the seedling of an idea that would one day flower as engineering for herbicide tolerance—which, when genetic technologies became commercial, would be the first widely used application.

Jaworski had supporters in Monsanto. In 1972, a senior vice president, Monte Throdahl, negotiated a research agreement with Harvard University to study cell culture. That year was time for a change of the guard at Monsanto, and John Hanley, who became president and chief executive of the company, saw the value of new directions. Jaworski recalls being invited to Hanley's office, where the big boss himself counseled him on how to present his ideas to Monsanto's board of directors.

Still, until the end of the decade, Monsanto was prepared only to wade—not dive—into the risky new waters. The company passed up the offer from Genentech, one of the early biotechnology start-ups, to market its human insulin, the first pharmaceutical product to spring from the science of recombining DNA. Monsanto resolved to develop expertise internally, and it moved aggressively in that direction in 1979 by hiring Howard Schneiderman, who became Jaworski's key ally.

Schneiderman raised the profile of biology at Monsanto. He was on the faculty at the University of California at Irvine, had a doctorate from Harvard, and was a member of the National Academy of Sciences. He was known for his infectious energy and his passion for science. In an unpublished biography, Schneiderman, who died in 1989 of leukemia, wrote a passage that illuminates a motivation of science:

"My greatest satisfaction is not in having power over people, but in having power over nature. There was a wonderful pleasure in understanding the rules of nature and, having understood them, making those rules work for me. That was exciting."

Recalling his early days at Monsanto, Schneiderman also wrote: "My biggest task in 1979 was convincing Monsanto management to move aggressively into biotechnology . . . At the outset, I identified Ernie Jaworski as a person I could trust and on whom I could depend."

THE M TEAM

Ernie Jaworski now had a healthy dose of Schneiderman's zeal and the institutional blessing to proceed. What he needed next was what Schneiderman had spotted in him: allies he could trust. He didn't consider raiding other chemical companies because, unlike Monsanto, they were not peering into the outer realms of biology. He began his search in universities. One of his letters went to Dr. Stephen G. Rogers, a microbial geneticist and assistant professor at Indiana University School of Medicine. Rogers had a brand-new job, but he couldn't resist a challenge. Jaworski and Schneiderman also persuaded David Tiemeier, a gene-cloning researcher at the University of California at Irvine, first to consult with and then to join the team.

Jaworski interviewed Robert Fraley, from the University of California at San Francisco, at Logan Airport in Boston while they were waiting for separate planes to fly to different conferences. Fraley, a postdoctoral fellow and an early expert in the field of DNA transfer into plants, was eager to join. Jaworski's final team member sought him out. Rob Horsch, like Jaworski, had studied tissue culture at the University of Saskatchewan. Horsch heard that Jaworski was recruiting, and he wrote a letter that won him a ticket to St. Louis for an interview.

"I must have convinced them that I would be good for the team because the weekend I got back, Ernie called and offered me the job," Horsch recalled.

From day one, the M Team's charge was figuring out how to transfer genes in plants, making the transfer work by whatever method and systems made sense. It was not the way research at a company like Monsanto, in search of definable products, typically is done. For Jaworski's team, it was like going back to school. The scientific questions came first, with little talk about the profits to follow. Rather than veiling their research in secrecy, Jaworski encouraged the team to work with scientists around the country and to publish in scientific journals.

Jaworski knew that the best scientists won recognition, as well as personal satisfaction, when they published. He also had been through Black Fridays at Monsanto, when researchers were pared from the staff. So if nothing came of this new experimentation or if axes fell, Jaworski wanted his young scientists equipped to find new jobs. It was valuable to Monsanto, too, because no one else at the company had expertise in these newly carved edges of science. "We could be feeding them a line, and they wouldn't even know it," Horsch told me. So writing papers and bouncing ideas at conferences or over drinks later were means of quality control that, Jaworski understood, helped to keep team members off the wrong tracks.

It was collegial and fun, but there was underlying tension. Any day, someone would invent what Monsanto planned to create. And the leaders would patent their invention—unless the M Team beat them.

If Jaworski's boys were to succeed on the fourth floor of "U" building, they had a few questions to answer correctly. First, was it possible to transfer new DNA into a living plant cell? Next, if you moved the gene in there, would it work? Then, would the modified cell grow into a plant that exhibited the trait that you had engineered? The transfer had to work, the gene had to work, and the selection had to work—or nothing worked. And they wouldn't know what had gone wrong: whether they had failed once, twice, or perhaps even in triplicate.

To transfer the DNA, the scientists used a bacterium—*Agrobacterium tumefaciens*—that produces a disease in plants, injecting a form of DNA called plasmids through the plant's outer surface into cells. Into those plasmids they had inserted a new gene, one that enabled plants to grow in the presence of the antibiotic kanamycin.

The Monsanto scientists were working both with a wild species of petunia and with tobacco, the world's most studied plant and the guinea pig of plant science because of all the money the tobacco industry poured in over the years. In the fall of 1982, the M Team constructed an elaborate experiment using the gene for antibiotic resistance. They also had several control groups of plants with no antibiotic gene. Horsch had tailored a "co-cultivation" tissue-culture system in which he combined the newly engineered *Agrobacterium* with plant cell protoplasts in a solution. Then he grew the protoplasts in petri dishes that contained both kanamycin, which would kill cells, and hormones to encourage their growth.

"I looked at these things three times a day, every day of the week, including Saturday and Sunday," Horsch recalled.

BREAKTHROUGH

What Horsch and the rest of the team began to see was heart-thumping stuff. The culture cells, hundreds of them, were rising from the petri plates. By now, the team knew failure well. So what they were seeing was unmistakable. Over and over again, the tiny petunia and tobacco plants were growing from a solution that should have killed them. The gene for antibiotic resistance worked.

There was no doubt. "It was unambiguous," Horsch recalled.

They had genetically engineered the first plants in a systematic way, a success that had eluded others with the temerity to accept the challenge. Other plants had been transformed, albeit with abnormalities, nonfunctioning foreign genes, or lacking the capacity to pass along the new traits to the progeny. At Monsanto, the team had engineered a plant with a working foreign gene in a way that did not damage the plant's physiology or that of its offspring. And it could be done again and again, as demonstrated by the swift adoption of the team's methods by hundreds of laboratories around the world.

Looking back, the scientists like to say that the reaction was quite subdued. After the breakthrough, the team felt the burden of an infinite amount of work still to be done. There was an entire plant kingdom out there awaiting them, thousands of genes to be understood. And, of course, there was a company that, before too long, would be looking for products and business devices to make a profit on its investment. Jaworski says that the team was so obsessed with what they were doing that the joy of the achievement was diluted over a series of successful events. "I don't think we had any big parties to celebrate," he said.

But the scientists knew they had achieved a breakthrough, and soon the world would know, too. There may not have been dancing in the streets of St. Louis, but in "U" Building not far from the Mississippi River, there was, in Horsch's words, "U-phoria."

Rob Horsch came from a family of scientific achievers. His father, James Robert Horsch, was an electrical engineer who developed equipment to assist the Apollo astronauts in hunting for water on the moon. Now the younger Horsch, by happenstance, prepared to bask in the acclaim for the M Team's achievement.

He had filled out an application to attend the Miami Winter Symposium, an annual conference that drew biologists from around the globe. So it was decided to use that forum to make the announcement of what the team had done. But Monsanto didn't have a speaking role. The night before the gathering, Horsch, data in hand, persuaded organizers to give him a microphone the next day. They made ten minutes for him on the program.

Horsch, who by then was thirty, was not widely known in the field of biology. He is diminutive in height and does not speak with a stentorian voice. What's more, accomplished scientists who had consulted for Monsanto—Mary Dell Chilton and Jeff Schell—sat on the same panel and had prepared announcements of their own related breakthroughs. Horsch was nervous. But when he spoke, the scientists heard clearly what he was saying. He had the data, in black and white, to back up his claim of success.

"Nobody said, 'no, I don't believe you.' It was just so striking, the result was just in your face. It had worked," Horsch said.

Afterward, he recalled, "the reception was amazing. People were excited; there was excitement in their voices. The nature of their questions was to understand how it was done. People came up and shook my hand. It was like 'who are you and where did you come from?' "

The *Wall Street Journal* made note of the achievement on its front page. But the story only mentioned petunias, not tobacco. Monsanto had decided that the negative connotation of tobacco would detract from what had been accomplished.

By the spring of 1983, the engineered cells had grown into flowering plants. The team used a system that became known in scientific circles as elegant to achieve more breakthroughs in DNA manipulation, and by early 1984, they had succeeded in showing that the plants passed on the antibiotic resistance. The team expanded its work to different genes and different plants. Following the petunia and tobacco successes, Monsanto engineered a tomato in 1985. In 1986, a canola plant was modified. The year 1987 was huge for the team: They genetically engineered cotton, soybeans, potatoes, flax, and alfalfa. Sugar beets were engineered in 1988 and then, in 1990, corn.

The team had grown to about forty researchers, but the accountants at Monsanto were nervous. The company still was not close to a commercial product. Genetic research had traditionally been given away by scientists, and no one had figured out how to make Monsanto's investment pay. Many in the company persisted in the belief that Monsanto's future remained with chemicals, even though the patent on Roundup was halfway through its life and no more blockbuster herbicides had been concocted.

On their side, the M Team had the backing of the company's new chief executive officer, Richard Mahoney, who had replaced John Hanley in 1984. But the company was entering the tricky, convergent waters of government regulation and public opinion. Both flowed into the next phase of experimentation: moving from the growth chamber into the real world. Instead of growing genetically engineered plants in a petri dish, they needed to sprout them in the dirt.

In 1986, I telephoned Jaworski from Washington to talk about the outdoor tests the company was planning to conduct with its genetically engineered tomatoes. There had been signals from regulators that approval might not go smoothly, and environmental activists were sizing up a new

One of the first anti-GMO demonstrations on American soil took place at Monsanto Company headquarters outside St. Louis in July 1998, and drew protesters from as far away as India and Japan. They were met by a sign that read: Greetings from Monsanto: Safety First—Cold Water Available." Protesters tore down the sign during their otherwise peaceful gathering.

ballpark. Jaworski was coy with me that day, saying only that his team intended to plant their tomatoes "in the Midwest." When I pressed him, he allowed that Monsanto had several sites where the modified tomatoes could grow, among them company-owned research farms near Troy, Missouri, and Jerseyville, Illinois.

"We're doing our darnedest to get ready," Jaworski told me.

SEEDLINGS OF PROTEST

In the late 1990s, Monsanto would be stunned when Europe exploded over its plans to sow genetically modified crops. Even proposals for field tests on tiny swaths of land mobilized communities and inspired midnight strikes by vandals when the crops were planted. Monsanto might have looked to the response in its own backyard more than a decade earlier to see what can happen when people aren't reassured or just plain feel left out.

In March 1986, the city council of St. Charles, just outside St. Louis, voted unanimously to oppose an outdoor genetic test at a farm beyond the city limits. Worries among the locals were kindled by the sudden surfacing of a debate in Washington about the wisdom of related experiments in California. How would government oversee the new life-changing wizardry? Which agency would take charge?

St. Charles was a community that lived with the legacy of earlier scientific experimentation: tons of contaminated soil from the bomb-making years. Council members, backed up by Mayor Melvin Wetter, worried that Monsanto would be operating with its revolutionary bacteria too close to homes and the town waterworks.

One of the council members, James A. Williams, said he was disturbed that city officials had not been given adequate information about the test. "We had to read it on the front page of the Sunday *Post-Dispatch*," the councilman said. It was an early episode of Monsanto alienating people with a public-relations strategy either too secret or too clever. It was the beginning, too, of my own protracted adversarial relationship with Monsanto, whose officials did not easily accept coverage in the hometown paper that was other than unfailingly supportive.

In this case, it was not tomatoes, potatoes, or petunias that troubled the locals; it was a genetically engineered microbe that Monsanto wanted to test in the earth. The company had engineered the *Bacillus thuringiensis,* or Bt, a naturally occurring bacterium favored by organic gardeners, into another bacteria that then would be coated on corn seeds. The aim was killing cutworms, a corn pest. The research was a forerunner of genetic engineering with Bt that would become common in plants in the late 1990s.

Had Monsanto been operating in a vacuum, there's a good chance that the locals would have trusted their famous chemical company. But now that the mysteries of moving genes between organisms were unlocked, new companies were popping up and racing one another for breakthroughs. St. Louisans began to read a steady diet of stories, many of them mine, about nervous federal regulators and experiments with strange names. Like "ice-minus."

A California company, Advanced Genetic Sciences, had received the first federal permit for what were ominously labeled "deliberate release" experiments. The company wanted to spray a genetically engineered organism that retards frost on strawberries in California's Salinas region. But the company landed in hot water with the Environmental Protection Agency when it was discovered that it already had conducted tests outdoors, on its roof, without federal approval.

In its beginning, genetic science had fallen into scandal. Rather than hearing about how genetic engineering could improve their lives, the public read about the government stripping scientists of permits and fining them for breaking the law. What's more, local officials in Monterey County, California, had imposed an outright ban on the ice-minus test, raising the specter of communities controlling science in their borders. "Those guys couldn't test-drive a Chevrolet in this county without a use permit," a county official remarked.

The EPA, which had taken control of engineered pesticides in the early federal regulatory scheme, was warning a fledgling industry to go slow. "We're sending a message that this is important and that it has to be done in a cautious and well-thought-out way in these early years so that the public has confidence in the process," John Moore, then an assistant EPA administrator, told me.

The EPA sent Monsanto an extra message: Government scientists reported finding flaws in laboratory tests the company had conducted in support of its proposal for the St. Charles tests of a gene-altered pesticide. The EPA wanted new safety studies to determine, among other things, if the bacteria could harm clams and other mollusks if they reached the Mississippi River. I quoted an aide to then-senator Al Gore as saying, "This is a company known for great scientific minds, and then they submit sloppy, substandard work like this. It really makes us wonder." It troubled the locals, too.

The EPA's diligence triggered an expanded public-relations drive by the company. In presentations to local officials, Monsanto began walking the fine line that always would prove uncomfortable: a chemical company touting genetic engineering as an antidote to the dangers of chemicals.

"If this is thwarted, we'll be forced to have chemical pesticides forever," company scientist Robert J. Kaufman told a gathering of St. Charles residents. "Not that chemicals are unsafe," he quickly added.

In the mid-1980s, Jeremy Rifkin had begun establishing his reputation as biotech provocateur. From his grandly named Foundation on Economic Trends in Washington, Rifkin raised questions about both the safety of tests and the social implications of the discoveries. "After thousands of years of engineering the cold remains of earth into utilities, human beings are now setting out to engineer the internal biology of living organisms. We are moving from the age of pyrotechnology to the age of biotechnology. Every new biotechnology discovery will affect the power that some people exercise over others," I quoted him as saying.

Rifkin could send Monsanto scientists into a rage. "He says genetic engineering is going to destroy the world; in fact, Jeremy Rifkin is going to destroy the world," Kaufman told one of my *Post-Dispatch* colleagues, Tim Poor. "Somebody ought to turn around and say, 'Is Jeremy Rifkin safe for the environment?'"

Rather than demonizing Rifkin as Monsanto sought to do, industry officials might have been better advised to learn ways to more effectively respond to him and to prepare communities for the attacks on genetic engineering that surely would be waged.

Monsanto was winning local converts for its St. Charles plan, but the clock was ticking. The company needed to plant the coated corn in the spring of 1986 or miss the growing season. Over $2 million had been sunk into this project. In early May, another California scientist, Steven E. Lindow, had won EPA approval for a different "ice-minus" experiment on a potato patch at a University of California farm three miles south of the Oregon border. Monsanto enjoyed being first, and now it looked as though Lindow might win the distinction of conducting the first licensed, open-air release of a genetically engineered bacteria.

St. Charles wouldn't back down. County Commissioner Rodger Parker issued a written statement saying that his office had been "inundated with phone calls pertaining to the Monsanto tests. Monsanto has been a good neighbor to St. Charles County for the past twenty-five years plus. We appreciate their willingness to work with us to resolve these legal questions."

On May 11, the Planning & Zoning Commission hand-delivered a nononsense letter to Monsanto. "Should you proceed with your project without the proper zoning, the county will have no choice but to strictly enforce its Zoning Order and Flood Plain Ordinance," it read.

A few days later, the EPA ordered even more laboratory tests before the permit could be granted. For Monsanto, it was the last straw. Facing opposition on two fronts, it dropped its experiment. In my story from Washington reporting the decision, a company spokesman, Gerard Ingenthron, called the reversal of fortunes "a temporary setback. Biotechnology is going to be a major industry, and we still plan to be a major participant in it."

HISTORIC TOMATOES

Ingenthron was right, but it would take another year to get back on track. By then, in the spring of 1987, the genie was bursting out of the bottle. Just as strong was the impulse to prevent the technology from leaping into the world. Advanced Genetic Sciences had squared its dealings with the EPA by agreeing to pay a fine for its premature plantings. On April 23, in rural Contra Costa County, fifty miles northeast of San Francisco, the company prepared to spray its Frostban bacteria on strawberry plants in what would be the first licensed, open-air testing of a genetically altered microorganism.

But fear also was taking root. That night, saboteurs struck, cutting through a chain-link fence and uprooting more than 80 percent of the 2,400 strawberry plants sprouted in the field for the experimental spraying. The attackers did not destroy the tiny plants; after removing them from the earth, they simply laid them where they were dug. The gentle handling made a daybreak replanting possible. Then plant pathologist Julie Lindemann and her crew set about making history.

At 6:46 A.M., clad in a white polylaminated anti-contamination suit adorned with the word "Frostbusters" and breathing through a respirator, Lindemann used a common hand-pump sprayer to douse the strawberries with an uncommon solution of a gene-altered *Pseudomonas bacteria*. Her precautions sent mixed signals to a public being told that genetic engineering is safe.

Six days later at Tutelake, California, in Modoc County, another EPA-sanctioned anti-frost experiment incurred a similar reception: Vandals uprooted most of 4,000 potato plants awaiting sprays of genetically engineered bacteria by University of California scientists. But the potato test also proceeded.

Back in southwestern Illinois, Ernie Jaworski's tomatoes had yet to be planted. But in the flat farmlands of Jersey County, there weren't many people to raise a ruckus over a tomato test with no firm date. Neither did the Agriculture Department object; unlike the EPA, it took a kid-gloves approach in regulating genetically engineered plants.

On June 2, 1987, Rob Horsch, Robb Fraley, and Stephen Rogers awakened before dawn to prepare for what might be another milestone in genetic engineering. On that day, they hoped to supervise history's first outdoor test of a genetically engineered plant. They drove across the Mississippi River and through Alton, Illinois, to Jersey County and Monsanto's research farm.

But federal approval had not yet arrived. Nor had the truck from St. Louis that had been loaded with modified tomato plants, the size that people plant in their spring gardens. The trio waited nervously. About the time they heard the truck, the telephone rang in a field office. Rogers grabbed the receiver. Go ahead, an Agriculture Department official said.

Some of the plants were engineered with the Bt gene for insect resistance, some for virus resistance, and some for resistance to Roundup, the company's famous herbicide. The tests were independent of each other, and they showed how far Jaworski's team had come in less than five years, when those petunia and tobacco plants were first engineered.

Into the ground the plants went, before somebody in Washington could reconsider.

A PRESIDENT HUMBLED

On April 27, 1999, the Monsanto team stood in a place far away from either "U" Building or a plowed field. It was the East Room of the White House into which Jaworski, Fraley, Horsch, and Rogers had been escorted by a military honor guard.

Around each of their necks, President Bill Clinton draped the National Medal of Technology, a hunk of bronze that hung from a red, white, and blue ribbon. The medals are awarded for accomplishments in science that have made "profound and lasting contributions to our economy and quality of life." They are awarded sparingly; on this day eight companies or individuals received them, among them Denton Cooley, the Texan who pioneered heart transplants.

"You have sought answers to questions that few Americans can even begin to understand and others that people ask, but can't answer," Clinton said.

"Your success in illuminating the hows and whys of the world and raising the quality of human existence has helped make the time in which we live perhaps the most exciting in human history. I am humbled by your achievements."

AFTERMATH

One year after Rob Horsch had humbled the president of the United States, I was lucky to find him. Monsanto's Agracetus subsidiary, which Horsch managed, sits on the edge of Madison, Wisconsin, west of the University of Wisconsin and the Capitol building. No signs directed me either to Monsanto or Agracetus; nothing on any building offered a clue as to where I was. Monsanto might have achieved epochal achievements, but when the new century arrived, it had become an object of global reproach. The company was not making approach easier for attackers, protesters, or even saboteurs. On this very day, demonstrations against the World Bank that included a noisy, anti-GMO faction were clogging Washington, the lead story on the hourly radio news.

Even close to home, Monsanto had plenty of detractors. When I visited Rob Horsch, the *Onion,* the wickedly funny newspaper of satire headquartered in Madison, published a "Biotech Foods" feature. To the uninitiated, it looked like any feature that might run in a tabloid. Alongside a graphic of a double helix sprouting from the ground was a box containing the results of a poll and these words: "Genetically modified fruits and vegetables are an increasingly common sight on supermarket shelves. What is their appeal?" The findings were peculiar, indeed:

- · 8%: Stay fresh, crisp and colorful for weeks, even in human digestive tract;
- · 12%: Prefer eating cabbage straight off the bone;
- · 11%: Monsanto cucumbers able to entrap and devour wildlife with their tentacles;
- · 15%: "Beefsteak Tomato" no longer just a fanciful name;
- · 13%: Crunchberries seen more wholesome when grown and picked from actual bush;
- · 10%: Enhanced carrots aid stranded castaways in spotting rescue ships;
- · 17%: Breakfast made more fun by eggs that scurry around chittering, "Oh, me! Oh, gracious!" while you whomp at them with frying pan;
- · 14%: Mr. Potato Head no longer an impossible dream.

The *Onion*'s tally came to one hundred percent, I tabulated as I drove. As I searched, I considered the burden of hiding out from the world each day, as Monsanto officials now were. I knew bright, savvy people at the company who wanted out but who had watched doors close when they identified their present employer. I wondered how long it might take the scorn and the insults to ebb.

The row of greenhouses appearing through my windshield suggested that I was roughly in the right place, but I still wasn't certain when I entered the vestibule of an unmarked building, where eight boxes with FedEx markings lay on the floor. I put on my glasses to read the labels and saw that they were addressed to the Agriculture Department in Washington and the sender's line read, Monsanto. Inside I encountered a locked door with a sign that read: GENETICALLY ENGINEERED ORGANISMS: AUTHORIZED PERSONNEL ONLY.

Rob Horsch is an amiable fellow, an earnest, unassuming man in his late forties by now whose thinning hair looks to be hiding a capacious brain. He reminded me of the guy on the dorm floor who was still gripping a textbook when the gang returned drunk at 2 A.M. I'd seen Rob on panels responding thoughtfully to aggressive questioners, and I'd felt sympathy for him.

On a table in his office was a copy of a newspaper article I'd written nearly fifteen years earlier, under the headline: "Genetic Engineering Poses Environmental Questions." In the middle of the story, which was spread over six columns on the front of the news analysis section, was a photo of Horsch peering into an electron microscope. The story began:

> WASHINGTON—Sometime in October, the U.S. Environmental Protection Agency will decide whether to let Advanced Genetic Sciences, of Greenwich, Conn., turn loose rat-tailed little bacteria, born in a laboratory glass, on a strawberry patch in California's Salinas Valley.

There was nothing wrong or unfair about the piece, except that it had nothing to do with the photo of Horsch accompanying it. Looking at the photo, I wondered why he had put the story on the table and how the explosion of adverse publicity had affected the scientists behind the world-changing discoveries. From my experience, scientists are constructed differently than the warriors in their companies who fight the public-relations and regulatory battles. Seldom are scientists public figures with developed resilience to attack. Even carrying on in obscurity, they could not have remained shielded from the global furor over the last few years. When I visited Ernie Jaworski in St. Louis, he told me that he had been "annoyed" by my coverage of what his achievements had produced. I asked Horsch if the attacks had exacted a toll on him.

"From an emotional standpoint, it's taken a lot of the fun and pleasure out of it," he said. "In another way, it has actually brought a richness by broadening perspectives on social issues. It's taken me into contact with other segments of society and other worldviews that I wouldn't have known. I found this fascinating, but the outcome is still pretty frustrating."

Like many scientists, Horsch was dismayed by the turn of events that had left the food biotechnology industry stumbling. "In retrospect, it doesn't seem like it could have turned out worse, so there must have been something different that could have been done," Horsch said, when I asked him what had gone wrong. "Having lived through it, it was a big surprise. A lot of the things that people asked for, I thought we did. For example, I went to Europe before any of the products were approved and talked to people about it, met with industry groups and met with consumer groups and answered their questions. I told people what was happening, asked what their concerns were, and they didn't get excited. It looked at the time like we were doing everything that people wanted us to do."

Technically, the Monsanto team that engineered the first plants was still intact. But Horsch had moved out of the laboratory and into the business end of the company, leading its sustainable-development business. He was working with small-acreage farmers in developing countries, looking for ways to help them conserve topsoil and achieve more yield by means of better seeds and wiser use of fertilizers and pesticides. His projects didn't involve genetically modified seeds because Monsanto didn't, as he put it, yet have relevant products.

As a businessman, Horsch has more time to think than during all those years spent in St. Louis genetically engineering plants. It was clear from our conversation that what happened since has troubled him on more levels than the personal.

"It's been distracting," he said. "It distracts energy and press attention away from the really big, serious issues. Big issues require thoughtful solutions because you have to enroll people in land use, water use, and biodiversity protection because they're not free. You have to make the case that they're going to impact your children worse if you don't do something now, and that's a thoughtful process. You have to enroll people in it and wrestle with it today to end up with a better tomorrow.

"The head butting, the misstatements, and the whole emotional circus is not conducive to the thoughtful process that is needed. That's a personal opinion."

3

A CRY9 SHAME:
WHEN THE RULES BROKE DOWN

Science cannot flourish except in an atmosphere of freedom, and
freedom cannot survive unless there is an honest facing of facts.
—Henry Wallace
Former Agriculture Secretary and Vice President

CRY9C SOUNDS LIKE an exotic chemical or perhaps a gas wafting from a shiny vat in a futuristic novel about frozen people awaiting immortality. In fact, Cry9C is just a tiny protein buried in the dirt. But it's a protein that won a sliver of immortality in the biotechnology debate by prompting reexamination of the United States government's patchwork regulations for genetically modified food.

Cry9C is one of a family of crystalline proteins, known as endotoxins, that come from *Bacillus thuringiensis,* Bt, a naturally occurring soil organism. Bt is deployed by gardeners and organic growers around the world for its insecticidal capabilities, thereby allowing them to avoid chemicals. Starting in the mid-1990s, genetic engineers harnessed Bt's power by splicing it into corn, potatoes, and cotton. Because the Cry proteins from Bt act the same way as insecticides, they sound dangerous. But the Environmental Protection Agency has welcomed them into the environment. They pose no "unreasonable adverse effects" on wildlife, surrounding plants and organisms in the soil, the Environmental Protection Agency reported in 2000.

Besides Cry9C, there's Cry1Ab delta-endotoxin, approved by the government for use in field corn, sweet corn, and popcorn, and Cry1Ac, used in several crops. There's Cry3A in potato plants. But Cry9C, which is approved solely for use in a corn called StarLink, has a distinction: Because of its unknown effects on humans, the EPA decreed that StarLink corn could be eaten only by animals, and not by humans.

"It is not possible for the agency to determine that there is a lack of allergenic potential from Cry9C based upon the available information," the EPA observed, noting that Cry9C does not break down at high temperatures or in stomach gastric juices, and therefore doesn't easily digest. That restriction did not deter Aventis CropScience, a subsidiary of Aventis S.A., of France, and StarLink's manufacturer. In the 2000 planting season, about three hundred thousand acres of StarLink corn were planted in the United States. Aventis told its distributors, who in turn were to tell farmers, that StarLink was meant to be kept separate from other hybrids and to be planted only with buffers to guard against cross-pollination.

In September 2000, an alliance of environmental and consumer advocates calling itself Genetically Engineered Food Alert made a discovery they announced to the world: Supermarkets were selling taco shells that contained Cry9C. The taco maker was no small operator; the activists identified the brand as Taco Bell Home Originals, a product line of Kraft Foods, which is a subsidiary of Philip Morris. The shells in question were sold in lots of twelve and eighteen and in a third package with sauce and seasoning.

The route of these taco shells to groceries offered a tutorial in modern food production. They were made for Kraft in Mexicali, Mexico, by Sabritas Mexicali, a wholly owned subsidiary of PepsiCo. Sabritas in turn had purchased its corn flour from a company called Azteca Milling, which had processed the flour in its mill in the United States, in Plainview, Texas.

Somewhere in the production chain, the StarLink—grown for animals, as is most corn in America—apparently became mixed with other hybrids categorized as yellow corn No. 2. Picture a row of country grain bins that receive corn from many farmers before shipping it elsewhere, perhaps to foreign lands. Here corn from neighboring farmers who planted different seeds becomes blended. Preserving that system of grain mixing along the production trail is one reason food producers and many farmers themselves are fighting against a global campaign to label modified food. They fear that it will hasten the arrival of a two-track commodity food system, requiring costly segregation and testing to keep conventionally grown grains apart from crops derived from genetic engineering.

Genetically Engineered Food Alert is an alliance of seven advocacy organizations: Center for Food Safety, Friends of the Earth, Institute for Agriculture and Trade Policy, National Environmental Trust, Organic Consumers Association, Pesticide Action Network North America, and the State Public Interest Research Groups. The activists had searched for a way to demonstrate the shortcomings of the government, particularly the Food and Drug Administration, in regulating genetically modified food. They knew from working with farmers that just a fraction of the corn grown in the United States is segregated. They knew, too, about StarLink. So they took a chance.

In his shopping cart at a Safeway store in Silver Spring, Maryland, just outside Washington, Larry Bohlen, of Friends of the Earth, piled taco shells and twenty-three products that contained corn. He then shipped them to Genetic I.D., a laboratory in Fairfield, Iowa, for DNA testing. In August, the results for the taco shells had come back positive for Cry9C. Realizing the potential of a faulty result—and the risk of taking on the American government and the country's biggest food producers—the advocates wanted retesting. The results came back the same.

At a news conference in Washington declaring the Cry9C discovery, Genetically Engineered Food Alert demanded a recall of the taco shells and condemned what they regarded as a permissive, largely voluntary method of regulating modified foods in the United States. "It gives the industry the idea that the government is not taking these risks seriously," said Jane Rissler, of the Union of Concerned Scientists. "This is not an imminent health hazard; this is a potential health hazard that is illegally on the market, and a government agency should take steps to end that illegal situation." Neither Rissler, a former EPA official, nor members of the alliance could identify a single taco eater who had taken ill or suffered an allergic reaction. Both Kraft and the FDA, which oversees food safety in the United States, said it would look into the matter.

Genetically Engineered Food Alert had already succeeded on two fronts: By winning press attention across the country, they planted seeds of doubt about the safety of the new wave of foods. A *New York Times* headline read: "Altered Corn Entering Human Food." Secondly, in identifying an unauthorized food in our diets, the advocates shined a light into apparent gaps in the U.S. regulatory system.

GOOD ENOUGH?

On paper, American's biotechnology regulation might seem less than reassuring to a mother weighing what to put on the table. The United States—unlike Europe, Japan, and Australia—does not require labeling of genetically engineered foods. Nor were there mechanisms in the regulatory system to find taco shells or any food that might contain genetic materials whose safety remained unproved. It took a public interest group spending over $7,000 to sponsor tests on ground-corn products.

In the United States there is no testing by the government of genetically altered food. Nor in America, biotech incubator of the world, was there mandatory, premarket safety testing by companies. In the case of food additives, the FDA requires tests for allergenicity and toxicity. But by 2001, five years after modified crops began sprouting widely in the United States, safety tests remained voluntary. This rule is consistent with a government decision years ago: In its regulations, the FDA does not differentiate between food derived from genetic modification and food from conventional means. Thus, it does not require additional testing for potential allergens that coalitions of consumers, scientists, and environmentalists have requested. Nor, adhering to the doctrine of "substantial equivalence," does the government require labels on food packaging advising shoppers that food is modified or contains gene-altered ingredients.

The skeptics worry that too few precautions are taken to avoid what Pioneer Hi-Bred, the seed giant, found in 1995 when it spliced a gene from a Brazil nut into soybeans. Fortunately, long before the protein-boosted beans went into the soil, the company had sponsored tests that discovered something scary: the gene transferred to soybeans was the source of Brazil nut allergies. The project was scrapped.

The incident proved a point argued by both sides in the food biotechnology debate: For the critics, it demonstrated that a gene-altered food might cause unexpected, even fatal problems. For the industry, it demonstrated that the normal course of research could forestall such problems.

Environmental regulations, as well as health precautions, are found wanting by the skeptics. The biotech-friendly Department of Agriculture oversees the thousands of field trials required before newly engineered crops reach the market. But it is largely an honor system in which companies notify the Agriculture Department's Animal and Plant Health Inspection Service of their research. There is no requirement to structure the experiments to assess environmental effects and no mandated tests to determine the impacts of GMOs that leave test sites. As a result, the government has passed up the opportunity to gather data on gene flow, impacts on the soil, and other potentially negative effects—data that could be used to refute arguments of ecological harm.

The EPA is regarded as the most vigilant of federal agencies. Even so, it lacks a strong, tailored risk assessment system with specific protocols for engineered plants, its own scientific advisory panel has concluded. One famous example is the monarch butterfly, whose larvae were killed by Bt pollen from modified corn in widely publicized laboratory tests at Cornell University in 1999. The EPA had required tests of the pollen on bumblebees and aquatic insects, but hadn't bothered to determine whether non-target insects, like the monarch, were susceptible.

Fish are another concern. The FDA stepped forth timidly to begin federal oversight of genetically engineered fish. The prospect of faster-growing salmon held appeal, but skeptics worried about the environmental risk when and if fish with disrupted hormone cycles escaped to breed in the wild. Yet by 2001, the FDA had developed no in-house expertise for evaluating risks and had not said when, or how, it would regulate.

States and citizens were not waiting for the government to protect them. In April of 2001, fearing contamination of Maryland waterways,

evidence for position — pro —

...or Parris Glendening signed a law banning the raising of genetically
...ered fish outside of enclosed ponds. A few weeks later, fishermen
and seafood industry representatives joined a coalition of more than sixty
consumer and environmental groups petitioning the government to declare a moratorium on marketing and importing gene-altered fish.

All this is chimerical, says Val Giddings, of the Biotechnology Industry Organization, which represents nine-hundred fifty companies and institutions with a stake in genetic engineering. "The challenge that has
been raised about testing not being mandatory comes under the category of true but irrelevant," he said, when I asked him to explain why
he thinks the federal regulatory apparatus is sound. "No company would
even dream of putting on the market a product that had not gone
through the consultation process [with the FDA]. The liability exposure
would be so great that it would be intolerable. No general counsel
would let them do that.

"The number of food allergies that are life-threatening is very small; a
dozen or so. It's possible now to screen data banks of gene sequences to
see if genes you're working with match up with allergens that are common, and companies routinely do this. Nobody is doing any work taking
anything from an allergen. That's one of the ironies. The truth is actually
the opposite from what the critics are alleging. At the University of California, researchers are trying to delete from peanuts the gene for a protein
that produces the allergen to those who are susceptible, like my two-year-
old son," he said.

In other words, biotech is the solution to a problem, not the cause.

"Some who are critical of biotechnology want the perfect to be the
enemy of the good enough," he continued. "The potential environmental
effects have been looked at exhaustively. They have been subjected to far
more scrutiny than conventional foods. The critics fail to take into account
one very important regulatory factor: that you have to frame things in a
relative process. The only way to do that is to compare risks and benefits
of conventional crops and the risks and benefits of engineered crops. When
you do this kind of analysis, engineered crops come out equal to, or less
than, conventional crops in the environmental risks that they pose.

"Farmers have three options. They can grow corn and let the corn

3 PROS/ evidence for public + farmers

borer eat it. They can use Bt corn, which may have some nontarget impact with beneficials like monarchs, if they are exposed, which is low probability. Or they can use conventional sprays. The EPA has reached a pretty definite conclusion as to which one is best," he said. He was right on that score: A day before, the EPA had issued a report that Bt technology had prevented the spraying of nearly a million gallons of insecticide in 1999, mainly on cotton fields.

"I have yet to hear an issue raised by the critics which is not covered by the Coordinated Framework," Giddings said.

"COORDINATED FRAMEWORK"

In the early days of the biotech era, the mid-1980s, science presented a regulatory challenge with which the United States government would still be wrestling in the next century. From the start, the government's goals conflicted: Regulate gently enough to encourage innovation—but energetically enough to reassure people they were being protected from new and untested forces.

I watched Monsanto, the leader in the laboratory race to design genetically modified food, deploy its Washington connections to persuade the government to adopt a set of regulatory guidelines. The company's motivation was rooted in sound judgment, political sensibilities that would lapse a few years later. Monsanto understood that government oversight was necessary to mollify consumers who might fear the heretofore unseen combination of gene-splicing and food. There was also a competitive reason for pushing new rules, albeit rules that weren't overbearing: They might discourage the proliferation of start-up companies and small-change operators lacking a Washington presence and unequipped for the expense and red tape of new testing requirements.

The industry was fortunate that biotechnology came along when it did—during the antiregulatory heyday of President Ronald Reagan who, in June 1986, signed the government's Coordinated Framework for Regulation of Biotechnology Policy. The fledging industry responded appreciatively. "I was struck by what I thought was due regard for commercial interests," Roger Kleese, vice president of Molecular Genetics, Inc., a Minnesota start-up, told me back then. A Monsanto spokesman, David Crosson, said his company was "very pleased that the emphasis is on the product and its use, and not just the fact that biotechnology is the production method."

The new rules spelled out a division of powers between federal agencies rather than putting any one agency in charge. By using existing laws, the framers of the regulatory scheme succeeded in keeping Congress out of the picture, thus keeping control and avoiding cumbersome new statutes. From the start, the task of applying existing laws to new science was tricky. Agencies groped for a rationale while competing for a piece of the action.

A memorandum in March 1983, written by an EPA attorney, Stanley Abramson, who was associate general counsel in the pesticides and toxic substances division, reflects a bureaucracy coming to grips with a new technology. "A life form, whether a virus, a bacterium or a more complex organism, is clearly comprised of chemical substances under the Toxic Substances Control Act because they are the combination of organic substances of particular molecular identities that occur in nature or that occur in whole as a result of chemical reaction."

Six months later, Abramson wrote another memorandum recommending that the proposed ice-minus genetic experiments to gird plants against frost be regulated not under the Toxic Substances Control Act but under the Federal Insecticide, Fungicide and Rodenticide Act, which people in government call FIFRA. "The extent to which the agency elects to regulate substances designated as pesticides under FIFRA is largely dependent upon the potential risks related to their use. Proponents of these biological ice-nucleation inhibitors claim that the substances will not have adverse effects on man or the environment. However, all of these safety claims must be fully assessed before the agency makes a determination on whether and to what extent to regulate."

The EPA aggressiveness worried other agencies, particularly the Agriculture Department, which did not warm up readily to regulating. The squabbling was amusing to Neil Harl, an Iowa State University economist who spent considerable time in the early 1980s studying agriculture biotechnology. "We were looking at every aspect of this under the sun. It was a turf battle, and it was clear that no one wanted to give a quarter of an inch. The FDA said it had charge of all food safety. The EPA was setting itself up for the environmental portion. And the USDA seemed to think it should be involved, even though it had become a cheerleader for the technology. It was clear that the administration did not want to go to Congress and put together a totally new system of regulation. They sort of reached a truce, a three-way split of the responsibility in a way that probably left no agency with the burden of knowing that they were going to be the one who had to look closely at all the new developments."

From the start, as Harl recalled, the question was whether the existing laws were adequate for genetically engineered products. In the spring of 1984, a new Domestic Policy Council Working Group set out to answer that question, concluding that no legislation was needed and searching federal statutes for existing laws that might apply.

On June 26, 1986, the White House finally published its Coordinated Framework, which parceled out the authority. The EPA was given the power to regulate pesticidal organisms. The Department of Agriculture was given broad authority to oversee genetically engineered plants and animals. Federal agencies were ordered to regulate genetic engineering products no differently than those achieved through traditional techniques. The FDA would adhere religiously to that prescription, which it spelled out in a benchmark 1992 policy statement reiterating that foods produced through DNA manipulation would not be regulated differently.

With concerns mounting in the new century, the FDA held public hearings, then announced in May of 2000 that it intended to strengthen its rules somewhat by requiring manufacturers to notify them and provide more documentation before sending modified foods to the market. The agency also asked companies to consider voluntary labeling.

"FDA's scientific review continues to show that all bioengineered foods sold here in the United States today are as safe as their nonbioengineered counterparts," Dr. Jane Henney, the government's commissioner of food and drugs, declared.

In January, 2001, four days before the Clinton administration limped out of Washington, the FDA issued a set of draft guidelines that did nothing to tamp down the smoldering resistance to GMOs in the United States.

The rules not only rejected mandatory labeling, they forbade companies to use the "no GMOs" label coming into vogue in whole foods markets. And instead of mandatory pre-market testing, the FDA prescribed a consultation process in which companies intending to sell modified food would need only to tell the government about it four months in advance.

The Biotechnology Industry Organization called the new rules a milestone, a signal that the food and biotech industries had emerged all but unscathed. "The U.S. regulatory system is a model around the world because it is grounded in science, not superstition or uninformed emotion," the nine-hundred-fifty-member association proclaimed.

Becky Goldburg, a senior scientist in the New York office of Environmental Defense, would dispute that the emotion she felt was uninformed. Goldburg was a member of a panel of experts appointed by the State Department and the European Union, which had recommended strong regulation and mandatory labeling. "Clearly," she said, "the FDA doesn't have a taste for regulating genetically engineering foods."

STUNG

Four days after the Cry9C news conference by Genetically Engineered Food Alert, I received a telephone call passing along a rumor: Kraft Foods planned to issue a recall of the Taco Bell shells. I doubted the veracity of what I was hearing. No food had ever been recalled because it had been genetically engineered.

But it was true. Kraft declared that Friday afternoon that they were recalling some 2.5 million boxes of the taco shells from homes and restaurants. "You should not eat any products containing Taco Bell taco shells," Kraft advised on its web site, adding, "at this point, there appears to be no evidence of adverse health effects." Consumers were told that they could return packages of the tacos for refunds. Also, the Taco Bell restaurant chain, which was owned by PepsiCo, announced that their version of the shells, which are made at the same plant in Mexicali, Mexico, were being tested for contamination.

Kraft said it had no knowledge of how the illegal ingredient made it into its corn. "The specifications for the corn Azteca purchased for the taco shells were confined solely to several varieties of conventional yellow corn, and did not include the StarLink corn," the company said.

The company was clearly embarrassed by the controversy swirling around a product that accounted for about $50 million in yearly sales. Stung both by the costly recall and the publicity it had generated, Kraft called on the government to tighten its rules for genetically engineered food. The company said that the government should stop approving engineered corn unless people—and not just animals—can eat it. Moreover, the food giant recommended a mandatory review of all plant biotech work rather than the voluntary system. And if there's not a fully validated testing procedure for identifying GMOs in foods—and often there isn't—those foods should not be approved for the market, Kraft said.

Strikingly, America's biggest food company was making the same arguments as critics demanding that government do a better job of regulating the technology before it became locked irretrievably in the food chain. Agriculture Secretary Dan Glickman said the incident proved that the federal agency needed to do a better job of segregating gene-spliced grains.

And the FDA? The agency called Kraft's recall "prudent," but rejected the suggestion that it had been slow to act. "This is not a case where we have illnesses or health problems," James Maryanski, the FDA's biotechnology food coordinator, told the *New York Times*'s Andy Pollack.

In the FDA's response, a contradiction lurked. The government said at the time that it knew of no illnesses or health problems from Cry9C. But if people don't know they're eating genetically modified food—and the FDA says that labeling isn't necessary—I wondered how people would have the information to report that they became nauseated or went into anaphylactic shock from food with altered DNA.

The biotech and food industries should have been pleased that the initial news stories about the recalled taco shells didn't mention the breakfast cereals, TV dinners, and other foods on supermarket shelves made from the blended yellow corn. Those stories were yet to come.

As the weeks passed in late 2000, two things became clear. One, far more of the unpermitted StarLink corn had reached the food supply than originally believed. Two, the damage to the industry was deeper than first feared.

After the initial voluntary recall of taco shells, the FDA ordered a government Class II recall, a category defined as "a situation in which the use of, or exposure to, a violative product may cause temporary or medically reversible adverse health effects." Companies ordered more products stripped from shelves; Mission Foods, the biggest manufacturer of tortilla products in the United States, alone recalled three hundred products. Within a month, the umbrella organization of environmental groups found Cry9C in a second and third brand of taco shells.

The situation was swelling into an emergency, albeit one muted by so much other news, chiefly the twisting conclusion of a too-close-to-call presidential campaign in the United States. The Centers for Disease Control reported that dozens of people who'd read the news accounts said they had been sickened by the gene-altered corn meant for animals. CDC tests later found no link between the illnesses and the tainted products. The giants in the food-processing trade, Archer Daniels Midland and ConAgra, started testing shipments for StarLink and rejecting tainted loads, raising the ire of farmers. Aventis CropScience, which had instructed its seed distributors to stop selling the tainted corn, disclosed that it couldn't account for about nine million bushels.

By November, the StarLink was detected in corn exports to Japan, American farmers' biggest export market and a country on the verge of a new labeling system because concerns about genetically modified food were running so high. Corn exports to Japan plunged, and South Korea, one of the largest importers of U.S. corn, banned new shipments.

Aventis and the industry hoped for a plan that might have all but ended the swirling debate. The company that triggered the furor in the first place asked the EPA to amend its StarLink approval to human consumption, which would have forestalled more recalls and diminished urgency all around. But in December of 2000, an EPA advisory panel refused StarLink its clean bill of health. The panel said that the corn had demonstrated "a medium likelihood" of triggering allergic reactions. But because so little of it was in the food supply, the panel added, consumers were unlikely to have developed sensitivities to it. Nonetheless, the panel concluded that the cases of at least seven and perhaps as many as fourteen of the people who had contacted the government with worries of adverse reactions merited further study. The EPA promised "an appropriate regulatory approach," meaning, among other things, that it was unlikely ever again to give the go-ahead to a biotech product meant just for animals.

Months after the Cry9C revelations, no proof existed of illnesses from eating StarLink. Nonetheless, the furor had cost the industry hundreds of millions of dollars, disclosed gaps in the regulation of modified foods, and demonstrated anew that the government doesn't know what people are eating. The regulatory debacle was felt by taxpayers, too: The government said that it would divert up to $20 million from a fund that helped farmers cope with natural disasters and instead give it to seed companies struggling to contain contaminated seed. The announcement marked the first federal bailout of producers damaged by biotechnology. The distribution breakdown evidenced monumental problems in separating grains, a harbinger of what is ahead in a rigidly enforced, two-track system of GMO and non-GMO—the direction in which the world is headed. It showed, too, the land mines that can explode at any moment along the genetic-engineering trail.

What began as a hunch by a Washington, D.C., environmentalist mushroomed into a costly headache for farmers and just about everyone who handles what they grow as it makes its way to grocery carts. It marked the biggest embarrassment suffered to date by the biotech industry and the newest challenge for those committed to building public confidence in a powerful new technology.

PLANTINGS TWO ⬅️

I arrive at the farm supply store this morning at seven-thirty, which is the time farmers in my part of the world do their business. It's before the arrival of the people in Chesapeake County who have realized their dream: a five-acre ranchette and a couple of llamas. I have come to buy herbicide for my field trials, in which I am testing my genetically engineered soybeans against conventional ones. Of course, I must buy Roundup to kill my weeds because that's what my pirated beans are engineered for; anything else would kill them. So I don't have many decisions to make. If I'm a farmer, I guess that my job is simpler.

I have always suspected that farmers buy chemicals not for their efficacy but for their names. Over the years, I have kept an eye on what was sold. There was Savage, Scourge, Scepter, Slam, and Snip, along with Trounce, Trooper, Trigger, Target, and Tomahawk. Also Rout, Retard, and Rugby alongside Punch and Prowl. There was Hercules and Big Daddy, Gallant, Chopper, Clobber, Fury and one that sounded like a computer game, Doom Command. I recall, too, Kill-All, Killex, and Kilmor. These names suggest to me that by the time farmers reach the field, they're prepared to do violence.

Or, perhaps, ready to ride off over the ridge, which is suggested by Stirrup, Stockade, and Stampede. (Or Stunt-Man, if they farm with flair.) In this cowboy nomenclature resides the most successful farm chemical ever, Roundup.

Roundup is a broad-spectrum herbicide, meaning that it kills a lot of different stuff. Roundup has the reputation of being less toxic to mammals than its predecessors while remaining deadly. Its chemical name is glyphosate, and billions of dollars worth of it have been sold since 1976. It has spelled the death of billions of tiny, unwanted plants and the birth of many bonuses for executives of Monsanto, which created glyphosate. Roundup is not just a success in the chemical world: It is a phenomenon of global business. Many years, it scored a 20 percent growth in revenues. I recall a financial analyst saying, "Roundup rules the world."

In 2001, Monsanto's golden goose dropped little goslings around the world. Monsanto's U.S. patent on glyphosate expired, prompting companies to decide if they can make their own version more cheaply than they can buy it from Monsanto. In any case, Roundup will live on, whether or not genetically modified crops flourish.

My only decision is how much of Everybody's Favorite Chemical to buy. I start with a quart. I stop to examine the back-mounted sprayers, like the kind worn by *campesinos* in Central America. I'd watch them working in their bare feet, spraying chemicals that could not be used in the United States because they were banned or unregistered.

That won't be the case in my backyard, where I will wear boots and everything is legal—except my seeds.

4

THE NEW GENE CAFÉ

For I dipped into the future, far as the human eyes could see, saw the
vision of the world and all the wonder that would be.
—Tennyson

Your scientists were so preoccupied with what they could that they
didn't stop to think whether they should.
—Ian Malcolm (as portrayed by Jeff Goldblum),
Jurassic Park; 1993

Is SHE NAKED?"

That is the question I have for my guide about the visage rising mer-maidlike from beneath us with an opaque globe in her embrace.

Three times I watched this woman at Monsanto Company's Beautiful Science exhibit, at Epcot at Walt Disney World Resort, and still I could not see what, if anything, she was wearing. Perhaps the film that featured her was blurrier than normal for it was reflected on black plastic rather than on the waters of an indoor pond. I had arrived in Orlando during renovation, it seems.

Maybe it was the New Age music, the darkness of the theater where the film was presented, or the spooky lights crawling up the polyester mesh glued on the screen. Perhaps I had spent too much time stumbling through genetically engineered plants or watching Colorado potato beetles succumb to plants engineered to kill them. Whatever the reason, I could not get this woman in focus.

"She's clothed," responds the Monsanto guide, Lance LeComb, matter-of-factly. "That's an issue that has never come up before. But it's a good question."

The Lady of the Pond, as the Monsanto interpreters call her, is a dependable way to indoctrinate males over a certain age. For children, there's the Tunnel of Bugs, at the entrance to the four-thousand-square-foot Beautiful Science exhibit, situated in Epcot's Innovention West Pavilion along the Road to Tomorrow. A man in a burgundy Monsanto shirt emblazoned with the patented Food, Health, and Hope logo with its bordered, eleven-leaf vine, hands plastic magnifiers to kids before they crawl inside to observe what squiggles behind glass in the earth.

Nearby, teens pause at interactive games perched on pastel stands. They might have been seeking Video Games of the Future, another attraction hereabouts, or the high-definition screens in the Ultimate Home Theater Experience. But they've landed at Beautiful Science, and they're learning what they're made of. When a girl in a yellow Mickey Mouse rain parka gives up, I slide over to play. On the screen is a blurry image of something microscopic that looks like a hairy forearm. "Is this a plant, an animal, or a human being?" a voice asks me.

"You're right. It's a soybean," I am told after I luckily slap the correct plastic knob with the palm of my hand.

"Want to try another one?" The new image looks like the underside of the moon. "Is this a plant, an animal, or a human being?" I strike a knob. "You're right. It's a human being." As my prize, I get an infomercial. The voice tells me that by genetically engineering a plant—rice—scientists are on the verge of providing more vitamin A to suffering people around the world.

On my way back to visit the Lady of the Pond, I notice a placard on a wall that reads: "All DNA in plants, animals and humans is made up of the same basic genetic code." I ask Lance LeComb something that is apparent. Here in Beautiful Science, is Monsanto attempting to drive home the message that because all genetic material on earth is similar, there should be little mystery and hence no fear in freely shifting DNA between organisms and species? He repeats the message I have been seeing and hearing.

"We're all made up of the same building blocks. All living organisms are," he says.

PASSION PLAY FOR A YOUNG POTATO

A funeral dirge played on accordion and washboard sounds as seventeen youthful actors dressed as human feet trudge across the gymnasium floor at St. Mary's College, in southern Maryland. They are players in a traveling morality play by Bread and Puppet Theater called *Passion Play for a Young Potato*.

"Dig the hole. Dig the hole. Dig the hole."

"Boogety, boogety, boogety."

Enter dragons and a wolf. There is shrieking. Feet scatter.

The potatoes that have been planted in the soil have begun to grow, as I can see from the masked actors rising from the floor.

An actor speaks.

"We at Monsanto, the proud creators of Agent Orange, the finest defoliant used in the Vietnam War, present the NewLeaf Potato, which has been genetically engineered to contain many of the necessary pesticides to provide the highest yield possible. That means that in the genetic code of all of the potatoes are pesticides to protect the potato from bugs. So when it's growing, it's not only growing into a potato, it's growing into a pesticide."

Bells sound and chanting begins.

"Food, Health, and Hope. Food, Health, and Hope. Food, Health, and Hope. Food, Health, and Hope."

The chants grow louder to "FOOD, HEALTH, AND HOPE, FOOD, HEALTH, AND HOPE, FOOD, HEALTH, AND HOPE."

"We at Monsanto," a single voice begins, "offer you food, health, and hope. The NewLeaf Superior Zone isn't a place, it's a state of mind."

My state of mind as I watch this play is slightly confused. I had driven to St. Mary's City, which is situated two hours southeast of Washington on a peninsula of land where the Potomac River meets the Chesapeake

Bay, after hearing that a traveling troop of actors would be presenting a play about genetic engineering. I had been expecting something else, perhaps naively, but what I discovered reaffirmed to me how deeply the issue of genetic engineering has penetrated not just science and agriculture but culture.

"Not every grower has superior confidence," the voice on the gymnasium floor continues. "It's a feeling you can only get from growing weed-free high-yielding NewLeaf Superior Potatoes. Only the NewLeaf Superior Potato gives you the feeling of confidence that you can get from ultimate weed control."

A whistle blows. "The preceding statement has been written by the Monsanto Company, all rights reserved. Transmission, reproduction, and distribution, including puppet shows, without express written permission of the Monsanto Company, IS STRICTLY PROHIBITED."

On stage, a pink dog is playing an accordion.

"Excuse me," a farmer asks. "Is this potato safe to eat?"

A voice for the company speaks. "We at Monsanto should not have to vouchsafe the safety of biotech food. Our interest is in selling as much of it as possible."

I recognize those words as having been uttered by a Monsanto official to a magazine writer. He told me later that the writer left out the rest of his quote, which was: "That is not our job. That is the job of the FDA. That's why we have independent regulatory agencies. The economic self-interest of companies is fairly obvious when they make claims about the health and safety."

Knock. Knock. Knock. "Excuse me, Mr. FDA, is this potato safe to eat?" the farmer asks.

"For the purpose of federal regulations, the NewLeaf Potato is not a food but a pesticide and therefore under the jurisdiction of the Environmental Protection Agency," the farmer is told.

"Knock. Knock. Knock. Excuse me, Mr. EPA, is this potato safe to eat?"

"We have found a reasonable certainty of no harm and therefore we have established no human tolerance for the product," the farmer hears.

Now the company speaks. "You are hereby licensed to grow Monsanto's NewLeaf Superior Potato for one generation only. The crop you water, tend, and harvest is yours, but also not yours. You may keep it and grow it, but it remains our property protected under U.S. Patent Number 5192535. If you should save and plant even one potato next year, you are subject to prosecution under United States federal law. You can go to jail."

There is a struggle on the gymnasium floor and dueling chants of "heave-ho" and "Food, Health, and Hope." From what I gather, the farmer and his potatoes break free of their bonds.

"The sun is risen," an actor says.

"Hallelujah, hallelujah, the sun is risen."

The lushness of genetically modified crops is displayed at Monsanto Company's Beautiful Science exhibit at Epcot at Walt Disney World Resort in Orlando.

GENETIC AGE VOCABULARY

Ganesh Kishore wants to add to *Webster's Unabridged* while he fattens scientific journals. The word he coined is "vitalin," which he defines as any of the compounds achieved through genetic engineering that help people maintain vitality. Kishore has sprinkled his new word in technical articles describing the coming array of nutritionally enhanced food that would revolutionize what we eat.

"I want to get my word popularized," says Kishore, the India-born scientist who was president of Monsanto's nutrition sector. "I think we ought to say that these are vitalins and this is what vitalins do. This is a new scientific lexicon being created here."

So new is the science of improving foods by DNA manipulation that we don't yet know what to call the results. There's "neutraceutical," which has a medicinal ring along with a bit of imprecision. There's "functional foods," which, besides conjuring up spoons of castor oil, applies these days to a slew of supermarket wonder eats that make promises bordering on quackery.

There's "value-added crops," a term with a sterile, mercantile ring that gives no clue as to who gets the value. Is it agribusiness and the middlemen, once again, who stand to benefit from crops bred with herbicide tolerance and insect resistance?

Had the ringleaders of biotechnology figured out how to better communicate the promise of their technology, they might have conquered first Europe and then the world. They might not have been whipped by the backlash that has frightened investors, narrowed product pipelines, and ignited public policy brushfires around the world. They might not be saddled with the expense of exhibitions like Beautiful Science at Epcot Center trying to win acceptance or have to endure scornful displays like *Passion Play for a Young Potato.*

They might have told us that coming to the fridge is a combination of food, synthetic nutrition, and medicine. What lies ahead, they might have said, is a revolution in how humans sustain life.

THE ROAD TO VITALINS

For most of history, our ancestors fed themselves by hunting animals and gathering wild plants. Only in the last eleven-thousand or so years did hunter-gatherers domesticate animals and plant seeds for food. Even then, some civilizations were thousands of years ahead of others. The lands first to learn how to systematically feed themselves are today, ironically, the developing countries: Southwest Asia (plants cultivated by around 8500 B.C.); China (by around 7500 B.C.); and southern Mexico and Central America (by around 3500 B.C.). Among developed nations, only the eastern United States (about 2500 B.C.) ranks among the top five areas of the world first to domesticate plants and animals. Western Europe, where the fiercest of the world's anti-GMO protests have raged, began growing its own food only when crops were brought in from elsewhere.

Anthropologists have traced how people evolved over thousands of years from hunter-gatherers, relying completely on wild food, to farmers, depending entirely on what they could raise. Along the way, they gradually domesticated wild plants into the foods common today. For instance, peas were domesticated by 8000 B.C., olives around 4000 B.C., strawberries during the Middle Ages and pecans not until the middle of the nineteenth century.

Genetic engineers insist that their manipulations in the waning years of the twentieth century merely extend the age-old tradition of painstaking improvement in the traits of food. Here is the crux of the debate. By moving genetic material between organisms and species, are scientists merely hastening the evolution of our vegetables and fruits, our fibers and oily seeds? Or are they redirecting evolution in ways whose outcomes are, at the least, unpredictable, and perhaps problematic?

Up to now, the corporate gene jockeys have aimed their technology at production ease, not at enhancing nutrition or taste. It can be argued that cost savings to farmers are beneficial to shoppers. Amid the environmental unknowns, there are benefits to the earth that can be counted: Reduction of chemical sprayings, especially insecticides, and the conservation of topsoil from no-till farming contribute to the common good.

But self-absorbed consumers—particularly aging baby boomers on the prowl for immortality—want something else: something for themselves: Researcher like Ganesh Kishore promise that these dividends from biotechnology are on the way.

"People continue to ask the question, what is in this for me? Is this a benefitless risk I'm taking for biotechnology?" Kishore says. "What we are aiming toward now is not about benefiting the farmer. It is not about benefiting a processor. It is about benefiting the consumers."

EATING FOR THE GENOME

In the dawn of the new millennium, research is converging at a three-way intersection of plant science, food science, and human biology. Scientists have known for nearly two decades how to move DNA between organisms, even species. The field of human nutrition has existed since the nineteenth century, and during that time researchers have figured out how food works in the body. We understand which vitamins, minerals, and compounds are essential for the diet, so we have the power to combat deficiencies and disease with food. We know, for instance, what to eat (and what not to eat) to stave off high blood pressure, heart disease, inflammatory ailments such as arthritis, even cancer.

Human biology, the third road at this convergence of sciences, could enable plant science and food science to make quantum leaps for human health. At least that's what researchers dream will come from deciphering the human genome.

Those mysteries about human genetic information were unraveled incrementally to the world in the late 1990s during an international race between the grand, generously funded Human Genome Project and the for-profit and less precise privateer Craig Venter and his Celera Genomics Group. Then, on February 12, 2001, in a declaration likened in gravity to Darwin's theories of evolution by natural selection and Mendel's pronouncements on the laws of genetics, the erstwhile competitors shared a stage in Washington to announce that they'd completed details of the entire human genetic map.

Assembled at the Capital Hilton Hotel in Washington by the journals *Nature* and *Science*—which would be publishing their papers in the coming days—the scientists brought the work of Watson and Crick full circle. They delivered surprises that spoke to both their successes and to the fallibility of prediction. They reported that a human gene can make two proteins or more, contrary to the one gene–one protein theory. They also declared that the human body contains just 30,000 or so genes, not 100,000, as the gene sequencers had been saying, and certainly not as many as 140,000, the number estimated by some researchers.

That revelation produced a humor competition between scientific teams that had labored in the lab to outdo one another. Eric Lander, head of genome research at Whitehead Institute for Biomedical Research and a leader in the publicly funded effort, said the revised human gene count offered a lesson in humility. "We have only twice as many genes as a fruit fly or a lowly nematode worm," he said.

Venter remarked that humans have roughly the same number of genes as corn. "Think of that the next time you eat corn," he said.

In April, 1953, Watson and Crick had written in the famously understated language of their landmark paper in *Nature* that the double helix structure "has novel features which are of considerable biological interest." On hand in Washington nearly forty-eight years later for the declaration that the human genome had been mapped, Watson declared it "a wonderful day." But the *eminence grise* of DNA scientists added that much work lay ahead. "All of us have different ways we want the knowledge to be used," he said.

The genome is the human body's entire DNA, arrayed in sets of twenty-three chromosomes in each cell. The DNA's double helix, recognized since the 1950s, is comprised of four nucleotides—adenine, thymine, cytosine, and guanine—expressed in the newly deciphered code by their first letters, A, T, C, and G. The genome mappers have managed to read the order of these nucleotides. Next, ambitious scientists say they will map the relationship between genes and disease. In the spring of 2000, I heard Craig Venter, founder of Celera Genomics Group and a pioneer in genome deciphering, tell a U.S. House of Representatives committee that in five or ten years, parents taking a baby home from the hospital will be handed a diskette containing their child's genetic code.

"It's putting power in individuals' hands over their own life, their own health," Venter said.

Biotechnology researchers see human health as their new frontier. The deciphered genetic code will enable scientists to determine causes of genetic diseases—weak links in the genetic code—and thus invent treatment to prevent the diseases. Similarly, unraveling plant genomes unlocks the secret door to greater precision in engineering the world's foods.

As the genetic engineers see it, food altered to contain more proteins, vitamins, or minerals is the prescription in a regimen of preventive health. Healthier food is the great benefit the life-science companies promise in their fifty-million-dollar-a-year information campaign begun in the spring of 2000, aired during the Summer Olympics at Sydney, and accelerated in the early days of the George W. Bush administration. It is the grand prize at the end of a tunnel of unknowns and the antidote to suspicions about their powerful new technology.

It's also a brave new world for the food industry: After working for a quarter century to take things out of what we eat—to produce food with less sugar, less fat, less cholesterol—scientists are now working to put things into food.

GOLDEN DISCOVERY

An early breakthrough in nutritional addition landed like food from the gods in the lap of an industry hungry for success. "Golden rice," so named for its pale yellow tint, is a genetically modified variety that produces extra levels of beta-carotene and related compounds that are converted in the human body to vitamin A. The engineering feat accomplished by Ingo Potrykus and his colleagues at the Swiss Federal Institute of Technology in Zurich has profound implications: One million children die every year because they are weakened by vitamin A deficiency.

When the news broke, the news media relished a fresh take on an aging story. Instead of seeing photos of angry protesters atop headlines reporting fear and loathing, readers saw accounts that trumpeted hope. "THIS RICE COULD SAVE A MILLION KIDS A YEAR," *Time* proclaimed in black bold-face on its cover in the summer of 2000. Alongside the words was a photo of the balding Potrykus amid a green wave of his ripening magic rice.

Potrykus and his colleagues overcame the primary challenge to genetic engineers seeking to add qualities to food. How, when you transfer a cluster of genes that amount to an entire biochemical pathway, do you control how they will express their traits? In this case, the scientists used genes from daffodils to produce beta-carotene inside the rice kernel. This blueprint for success is the one microbiologists will roughly follow as they pursue more such dividends: Genes from the selected plant or animals or other organisms are inserted along with other segments of DNA, called promoters, into small circles of DNA that occur inside a commonly used bacterium, *Agrobacterium tumefaciens*. The newly constituted bacteria are mixed in a petri dish with rice embryos, which they "infect," and in so doing transfer the genes that carry the instructions for making beta-carotene.

Traditional plant breeders had not mastered the crosses. And beta-carotene enhancement in rice was not a priority of the multinationals, even though three billion people count rice as their major staple. In Britain, a similar success was announced in the summer of 2000 when researchers at the University of London working with Zeneca Plant Sciences announced that they had tripled the amount of beta-carotene in tomatoes.

But rice is what feeds people, and Potrykus's early research was just the sort of project that would capture the eye of the Rockefeller Foundation, which in 1993 propelled his work with a hundred-thousand-dollar grant. What followed was a multiyear, 2.6-million-dollar project that also drew European backers.

There remain many obstacles, and perhaps many years, before bags of yellow rice reach the poor farmers for whom they are meant. Foremost is the thicket of patents on the genetic materials and the processes used by the scientific team. London-based AstraZeneca, which held exclusive license to one of the key genes, agreed in the spring of 2000 to relinquish claims. Monsanto followed suit.

At the Rockefeller Foundation, the 3.5-billion-dollar New York-based philanthropy, the beta-carotene success was a champagne-popping event. Long a promoter of genetic technologies to feed the world, the foundation had sunk over $100 million into research to produce modified varieties of rice. Amid the exultation, the foundation's Gary Toeniessen warned that other hurdles remained: probably two dozen or more patents held not just by companies but also by profit-hungry universities.

"But our feeling is that it's a doable task. I'm more confident that the companies will agree to help rather than the universities. I think that companies are getting better in the sense that they are more willing to enter into these negotiations. To be frank, I think that's in part due to all of the poor public relations they've been getting about GMOs and corporate takeover of the food system. They're receptive to the opportunities where they can demonstrate that they are willing to have a public interest," Toenniessen told me.

A "THIRD DIMENSION"

For the biotech industry, yellow rice is a matter of public relations, not commerce. Tailoring food for the developing world amounts to only a small line in multinationals' research and development budgets. The lion's share is directed to the tables of the industrialized North, bringing crops grown with genes for herbicide tolerance and insect resistance, meat engineered to contain less fat, and fish bred with a gene to speed growth.

The list of foods on which engineering experiments have been done is long: Besides the staples, companies and universities have notified the United States government of field trials on apples, coffee, cranberries, cucumbers, eggplants, melons, oats, onions, peas, pineapples, plums, raspberries, squash, strawberries, sugarcane, sweet potatoes, walnuts, and watermelons.

Potentially, virtually everything we eat can be derived from genetic modification, if the public is willing. What we drink, too: our milk from cows given genetically engineered growth hormone; our soft drinks with syrup from engineered corn; our beer from modified barley; even our coffee. In the summer of 2000, researcher Alan Crozier of the University of Glasgow declared that he had isolated one of the genes that both tea and coffee plants need to make caffeine. "All you're doing is switching off caffeine—not the compounds that affect flavor and aroma. It should taste like the real thing," he said.

At Cornell University, researchers announced in the spring of 2000 the development of a potato that can immunize patients against the primary cause of viral diarrhea. "This may very well become the first commercially available edible vaccine," asserted Charles Arntzen, leader of the research team.

In the industry's lavish information campaign, a web site trumpeted research to create "numerous food products" to combat heart disease and cancer. The advancing science may dictate not only what we eat but how we dress for dinner and how we arrived at the restaurant. On the Internet, companies pointed proudly to a future where clothing derived from corn and engineered crops "may help reduce our dependence on oil and natural gas and could reduce water and energy use by as much as fifty percent."

The "designer foods" on the drawing boards solve more than one corporate problem. Besides giving consumers something tangible, some of these boosted foods would enable companies to avoid a roiling debate about food labeling and choice. Companies would eagerly label these products as a pathway to health even if the Food and Drug Administration isn't compelled to order labeling of modified food with medicinal properties.

For instance, in 2001, Monsanto was working on a genetically engineered canola that it claims increases by 10 percent or more the level of stearidonic oil, which has inhibited colon cancer in tests on laboratory animals. As part of its focus on cardiovascular health, the company also has in clinical trials a food product aimed at reducing blood pressure.

Monsanto's Ganesh Kishore spoke of the imminent arrival of vitality-enhancing compounds encased in what he called unit-dosing foods, such as mints or single-sized servings of orange juice, milk, or even soup. His company is working to identify food compounds that can inhibit the onset of arthritis, osteoporosis, and even cancer. Similarly, compounds that activate these afflictions could be removed. He explained the thinking behind the research:

"Molecules of food can affect metabolic processes in the body and contribute to an acceleration or deceleration of these processes and thus affect the rate at which our cells, tissues, organs, and the body ages or deteriorates into disease conditions. This is the third dimension of activity associated with nutrients or components of our food. We are trying to understand these activities of food at a molecular level. . . . Our hypothesis is that such molecules can be effectively used to manage a range of chronic health conditions that plagues the human race today."

The vitalin-minter continued: "In the long run, we see these molecules being part of the diet and exercise regimen. If you had blood pressure slightly higher than normal, and I was to tell you that you could treat yourself by eating five servings of a certain food each day and avoiding meat, you might not want to go through that regimen. But it might appeal to you if I said you can eat meat in moderation and every day eat one of these tablets that won't cost you an arm and a leg. Maybe twenty dollars a month," he said.

PROTEIN PARADE

At its Protein Technologies, Inc., at Checkerboard Square in St. Louis, DuPont operates on both sides of the genetic-engineering debate. The company sells "identity preserved" GMO-free soybeans in Europe to exploit consumers' aversion to genetic changes. All the while, Protein Technologies invests in research aimed at selling more genetically modified soybeans in the seventy countries where it does business.

An early goal of DuPont was turning the healthful soybean into something people want to eat. One of the reasons soybean milk can seem unappealing is the enzymes in soybeans. But by genetically changing the content of oil, Protein Technologies researchers hope to remove that beany taste. Likewise, research aims at a bean that fortifies itself so as not to need additions of methionine, an unappetizing amino acid.

"From our standpoint, taste is number one, number two, and number three. You cannot market mainstream products without taste for consumers. When you want a refreshing beverage every day, you don't want it to be fatiguing to drink. You don't want it to taste like medicine. Biotechnology is allowing us to improve this," said Greg Castle, executive vice president for Protein Technologies' marketing.

He made his point again: "Our whole emphasis is better tasting foods; soy protein already is very high in protein quality. Soybeans can sustain the growth of a human being; you can't make it more nutritious. All you can do is make it taste better, creamier, and less like soybeans."

Castle sees this newly palatable soybean finding a home with all ages: from older people in assisted care, where dietitians work on providing the best nutrition, to the most active set—"high-school athletes, college athletes, weekend warriors, forty-year-old women who exercise three times a week. You can customize nutrition so as not to get muscle fatigue and you rebound more quickly," Castle said.

DuPont is working to alter texture as well as taste. By modifying the soybean and its oil, the company hopes to avoid the chemical additives needed to keep soybean-derived margarine in a solid stick.

Evidence of new research is reflected in records at the Department of Agriculture, where field trials on genetically engineered plants are recorded. Here, alongside descriptions of tests for the "input" traits of insect resistance and herbicide tolerance, more notifications began showing up in 2000 studying the "output" traits that change the constituents of food. With the cryptic denotation of "PQ"—which stands for product quality—companies offer glimpses of products of the future. Monsanto, for instance, reported efforts to increase levels of amino acids in corn along with experiments to improve oils from soybeans and canola.

Arnold Foudin, a biotech expert in the U.S. Department of Agriculture, has witnessed the evolution to the point where roughly one-third of field trials are aimed at adding value to crops. Much of the testing has been directed toward changes in animal feed, particularly corn. Foudin's division in the government's Animal and Plant Health Inspection Service, which oversees the field trials, had just received notifications of tests on "new and more exotic proteins and enzymes" in plants, he told me in the summer of 2000. There were four experiments aimed at growing vaccines in corn, so rather than administering vaccines via syringe, farmers soon may be treating their livestock with feed. Still more experiments under Foudin's oversight have been designed to help animals digest food more efficiently, one way of combating the nutrient damage to rivers and lakes from animal manure. Foudin sees this research as a precursor to developing a broad range of altered foods for humans.

"Rather than fight the public-relations problems of 'Frankenfoods,' companies are feeding their new crops to livestock," Foudin said. "The companies are probably correct to conclude at this point that if humans aren't consuming it, they won't face as much controversy. Feeding it to animals first is a normal progression in the way things in agriculture have worked. We cloned Dolly before we will clone a person. If we modify corn for animals, there's less consequences than, say, in so-called 'Frankenberries' for people. It's not written down, but that's the way it is. But really, what's the difference between feeding humans and other higher mammals? There is not a lot of difference in nutritional requirements; we could probably be put on the same diet as chimps and we'd do really well. For companies, it's a lot easier if you can demonstrate the safety when livestock is eating all these improved feeds. And then you go to your next step—human consumption," he said.

What Foudin is seeing, he says, are the early stages of developing foods resembling natural medicines. He offers the genetic engineers a goal to shoot at: "If you came up with, say, hormone replacement therapy for women that could be administered in two chocolate cookies, it would be a billion-dollar product. That's where we are heading; we've gone full circle, and we're coming back to natural medicines."

That's not good medicine to everybody.

JEREMY RIFKIN

To Jeremy Rifkin, as might be expected, all this talk about another wave of genetically modified products is folly. "I had a meeting with a top guy in a global company last week in Europe. And he said to me, 'Jeremy, it's over,'" Rifkin said, exhibiting his dual tendencies to stress his contacts and to forecast the impending death of a technology he reviles.

Rifkin, author, lecturer, and irritant nonpareil of the biotechnology industry from its inception, was surrounded by shelves of books he has written as he lectured me in his Washington office at his Foundation on Economic Trends. He had just arrived from Europe, where he was spending about half of his time hawking books, speaking for pay, and operating as *agent provocateur.*

He has written or co-written fifteen books, the latest being *The Age of Access: The New Culture of Hypercapitalism Where All of Life is a Paid-For Experience.* Before that, Rifkin's titles included *Biotech Century, Beyond Beef, Biosphere Politics, Time Wars,* and *Declaration of a Heretic,* which has these words alongside his photo on the book jacket: "Look closely at this man. Many leading scientists and businessmen greet the mere mention of his name with scorn, ridicule and hostility."

I found that out early, in 1986, after I wrote a newspaper profile on Rifkin that began: "From a tiny office packed with books he wrote and a clock that sometimes stops, Jeremy Rifkin labors to stuff the genie of genetic engineering back in the bottle." The article provoked a letter to the editor of the *St. Louis Post-Dispatch* complaining that I had given "authoritative prominence to Jeremy Rifkin, the self-styled expert foe of biotechnology, as he delivered his familiar complaints." The letter recalled a description of Rifkin as a "carnival pitchman" and quoted a columnist as saying, "The only danger to humanity lies in continuing to listen to Rifkin." I hadn't paid much attention until I got to the signature of the letter's author—Richard J. Mahoney, chairman and CEO of Monsanto.

For years, the biotech industry in the United States wished that Rifkin would just go away. He has, in a sense. But he may be causing the industry more problems than ever by stoking anti-GMO rebellion among government leaders in Europe and Latin America. When we met, he had just returned from Switzerland, where he appeared at the request of the Swiss government on a panel that included Nobel laureates. One of his points, as he related it to me, was vintage Rifkin: "I said, tell me you have a risk assessment science, and I will walk away. Tell me you have even the rudiments of a risk assessment science. They don't have one, because there's no computer modeling available on the horizon that can deal with the number of variables when you place any given gene and then you watch it flow in every ecosystem in the world."

As reported in Europe, Rifkin was influential in 1998 in persuading the French government to push for a *de facto* moratorium in European Commission approvals of new genetically modified crops. Early in this new century, the moratorium holds. These days, he's lecturing gatherings of European executives about economics. He argues, for instance, that the "market" exchange of property between sellers and buyers is going obsolete. Business as we knew it, he says, is being replaced by a system of "as you need it" access to virtually every service through vast commercial networks. What he describes, it occurs to me, sounds a lot like Monsanto's marketing of its genetic technologies, like Roundup Ready seeds, to farmers through rival life-science companies.

Rifkin, a Chicago native, also lectures at the University of Pennsylvania, where he was president of his graduating class in 1967. Then he was an antiwar protester embarked on a path of radicalism. In 1970, after graduate school at Tufts University, where he got a master's degree in international affairs, Rifkin collaborated with Jane Fonda in organizing congressional hearings for Vietnam veterans to tell their stories. After that, he was instrumental in organizing the People's Bicentennial Commission, the antiestablishment culture promoter of the 1970s.

Since 1976, when he cowrote *Who Should Play God,* Rifkin has focused on genetic engineering. With the persistence of a pit bull, he writes, organizes, lectures, and sues about what he sees as the inherent dangers and inequities of genetic sciences.

I hadn't seen Rifkin, who is in his mid-fifties, for eight years. I wanted to see if he had mellowed, whether he was ready to say that he was wrong about the bounty of biotechnology and to toast the inventors of golden rice. I also wanted a glimpse of what the biotech companies might expect from the opposition in the coming years. It didn't take long to get my questions answered.

"What I'm saying to political parties in Europe and NGOs and companies is that there are more problems with the second generation than with the first. When you code for a protein or a drug or a vitamin, it means that you're turning millions of acres of land all over the world into chemical factories," Rifkin began.

"There hasn't been a single discussion in any parliament of the world or in any scientific academy about the environmental implications of what they are talking about. What happens to foraging birds and insects and microbes and other animals when they come in contact with millions of acres of plants that are coding proteins for vaccines, vitamins, and chemicals? The problems they are going to have are a potential nightmare. And what about liability? And the health implications? A lot of these proteins are going to be innocuous. But you're going to get proteins that will be toxic, and the question is, who's it going to be toxic to? It might be one person in ten thousand.

"If I were a company in this, I wouldn't want to deal with an introduction where there's no long-term risk assessment, no insurance company to take care of long-term losses, and the health implications are unknowable. And finally, the market isn't there."

Surely, I thought, Rifkin would tip his cap to the discovery of the beta-carotene-enhanced rice. I was wrong again. "There are very cheap ways with vitamin A deficiency. This quick fix might be good propaganda for the companies. But I would ask them how local populations are going to be able to afford it. I think it's a bit disingenuous."

Soon, my conversation with the foremost critic of genetically modified food had run full circle. "If I were making a bet," Rifkin said, "I'd say that genetic foods will be looked back on as one of the great failures in the introduction of a new commercial technology."

No matter how good and good for you is the food inside, there are protesters outside as the New Gene Café opens its doors.

5

WINGS OF A BUTTERFLY: MARTINA VERSUS MARGARET

It is necessary for science to defend itself, first against attacks, and second against the consequences of its own success.
—Henry Wallace, 1933.

AFTER NIBBLING ON milkweed dusted with the pollen of gene-altered corn, the three-day-old monarch caterpillars crawled more slowly than usual, reflecting what was happening inside their bodies. Soon, their brilliant orange markings started to fade. Toward the end of the four-day experiment, they fastened themselves in a death grip to the plant that constitutes their sole sustenance in life and their principal food when they become butterflies. Then they turned black and began to rot.

"Caterpillars are just mouths and stomachs. All they're doing at this stage is eating and growing and shitting. These ate happily, they got sick and miserable, and then they stopped eating. They didn't eat that much of the leaves, but they ate enough that their guts were essentially ripped out and they became mighty sick animals," recalled biologist Linda Raynor, of Cornell University, who supplied the monarchs that became martyrs in the global debate over genetic engineering.

In the winter of 1999, a Cornell colleague of Raynor's, John Losey, an entomologist, was studying how the toxin from corn engineered with *Bacillus thuringiensis* (Bt) was affecting the intended victim, the European corn borer, a pest that causes over one billion dollars in damages every year to crops in the United States. Raynor, who tends a fourteen-year-old colony of the familiar, black-and-orange monarchs, joined with Losey and his assistant, Maureen Carter, in one of the early experiments to find out whether the expanding use of corn engineered with Bt against corn borers posed harm to another Lepidoptera, the monarch butterfly. Little did the researchers realize what they were stirring up in a Comstock Hall greenhouse in Ithaca, New York.

In the results of their experiment published in the journal *Nature,* the researchers said that 44 percent of the caterpillars that had eaten from the Bt-laced leaves had succumbed. But every caterpillar in the experiment's control group survived after dining on milkweed with the non-GMO pollen from conventional corn. Never mind that just eleven caterpillars died. Or that the pollen might not have been spread uniformly on the leaves and that they had run out of the pollen, preventing them from broadening the scope of the project. Or that laboratory experiments don't approximate the real-life field conditions that test the monarch's survivability. Even with those limitations, the Cornell monarch study became a flash point in a widening debate over genetic engineering. It lent European nations a fresh rationale for resisting GMOs. And it triggered a furious debunking effort by the biotechnology industry.

Under the chaos theory of events, a butterfly flapping its wings could set in motion the forces that trigger a hurricane elsewhere on earth. The butterflies of Cornell generated a hurricane-force debate, one that was not anticipated by Linda Raynor, monarch landlady. In experimenting with monarchs, the scientists had chosen a universal symbol of nature: the butterfly that is studied by children, released at weddings, photographed for paint store brochures for its brilliant hues, and admired by people of all ages for its beauty and its remarkable, three-thousand-mile migrations. At Cornell the researchers didn't fully grasp at the time that they were testing the world's most provocative new technology on the Bambi of the insect world.

"It was amazing," Linda Raynor recalled more than a year later. "I've never really done work that a lot of people cared about before, and this ended up shaking the whole biotechnology world. We knew it was interesting, that it was a hot potato. But we never expected that the monarch would become a mascot. We never expected the international furor that it would cause. The industry told us that they would attack us as scientists, and they did."

On the day the study was released, the Biotechnology Industry Organization scrambled to refute it, arguing in a news release that threats to the monarch are minimal: "Based on known migration behavior, even in those regions in which corn and monarchs co-habitate, only a small portion of the monarch population will be present when corn is shedding pollen," the release said.

In November 1999, at the industry's so-called Monarch Butterfly Symposium at Rosemont, Illinois, near O'Hare Airport, scientists paid to conduct studies further disputed the Cornell findings. An industry that had fumbled the public relations of biotechnology orchestrated a master stroke. Calling itself the Agricultural Biotechnology Stewardship Working Group, the industry organized a conference call between reporters and its scientists that cemented the story line before the symposium took place. A Canadian scientist, Mark Sears, of the University of Guelph, had found that 90 percent of the pollen grains travel fewer than five yards from the field and virtually all of it landed within ten yards. And John Foster, of the University of Nebraska, asserted that "habitat destruction, mowing and spraying rights of way with chemicals" endanger monarchs more than Bt. Asked if he had been swayed by his funding from the industry, Foster replied: "They don't give me enough money to buy my opinions."

The Cornell experiments triggered a series of publicly funded experiments in a technology that had relied on industry testing to demonstrate the safety of products. In the explosion of publicity following the work at Cornell, many people were troubled to learn that few studies before Bt corn was growing had measured potential damage to the four thousand species of Lepidoptera—butterflies and moths.

In the summer of 2000, another study, this one at Iowa State University, found that one in five monarch larvae died after being exposed to the toxic corn pollen for two days. And three days later, more than half of the caterpillars in the study were dead. Even so, the U.S. Environmental

Protection Agency deemed these studies "not useful" for assessing risks of Bt crops without more field studies.

In the revisionist look at biotechnology now under way, potential environmental consequences lay at the core of the debate. Increasingly, environmental harm—along with potential human problems such as allergenic reactions—have become part of an equation in which the public measures the risk of biotechnology against its rewards. And with Bt there is a reward—the reduction of spraying of insecticides which, like the Bt, also threaten monarchs.

As promise is balanced against peril, some scientists say aye and others nay.

Among those scientists are Martina McGloughlin and Margaret Mellon, whose views contrast the promise and the perils of genetic engineering.

MARTINA: BENEFITS OF GM FOOD "FAR OUTWEIGH THE COSTS"

The visitor cast his gaze around the farm in western Ireland town of Hepford, near Galway, and inquired about Jack McGloughlin's sons and the availability of home-grown help to milk the cows and till the soil. He was told that the McGloughlin offspring consisted of three daughters—Martina, Nuala, and Patricia.

"Well," the visitor said, pausing, "I hope you have a good dog, then."

The McGloughlin girls might as well have been boys, considering their regimen. They planted the potatoes in the spring and, all summer long, were dispatched to the fields like human herbicides, yanking the weeds from the patch so that they didn't divert the soil nutrients needed for the spuds to swell fat and solid. When the Irish switched to the metric system, Jack McGloughlin took bright little Martina with him to the market so he wouldn't get cheated selling his cattle.

Nuala later took over the farm, and Patricia works for the phone company. Martina charted a different route, through the fields of academia, a path that landed her far away in twin positions as director of the biotechnology program at the University of California at Davis and director of the Life Sciences Program for the entire University of California system.

With her comeliness, her easy laugh, and persuasive manner, Martina McGloughlin, who was born in 1959, made a persuasive new millennium spokesperson for an industry that needed good press. By writing op-ed articles for newspapers and traveling to biotech conferences coast to coast, she took up a challenge that many scientists had shunned. McGloughlin's background on the farm plays a leading role.

She told me her story.

"When you grow up on a farm, every spring you cut the eyes out of the potatoes. Each eye will give you a new plant; each of those is an embryo from the tubers. So we knew all about plant reproduction before we went to school. Likewise, we knew about animal reproduction. In Ireland, when the A-I men came, that didn't mean artificial intelligence. It meant artificial insemination.

"Most of what we ate, we grew. It's not like you couldn't go to the supermarket, but we didn't have the money to do it. So you're far more in touch with the soil. None of us would have grown up thinking that milk came from the supermarket, whereas a lot of American kids think they're drinking supermarket milk. There's a disconnect between agriculture in the United States. There isn't an appreciation for the integrity of agriculture and how tough it is for a farmer to survive. I see my sister, who took over for my father. A lot of it is living hand to mouth. Mad Cow Disease in England destroyed her profit because what she has are steers."

Listening to McGloughlin, I got a keener sense of the roots of the anti-GMOs sentiments in Europe that had all but halted the worldwide spread of the technology that she viewed with hope. In Ireland, she said, the Great Famine of 150 years ago remains recent history. Perhaps that is why, I thought as we spoke, when she returns as a celebrity to her native land she has been confronted by protesters fearful of mass alteration of their food. That was the case in 1999, when McGloughlin took part in a conference at Trinity College in Dublin, alongside luminaries such as Ian Wilmut, the Scottish scientist of Dolly-cloning fame. "Why," she was asked from the audience "are you working for the evil empire"—Monsanto?

"They said to me, 'you were brought up in this country with a sound regard for agriculture. How could you do this to us?' "

As she has showed in forums around the United States, McGloughlin is not shy in responding to her accusers.

"I said the reason I'm doing what I'm doing is precisely because I struggled in my youth. I spent my springs on my hands and knees planting potatoes, and I'd spend the rest of the summer down weeding. I was up each morning milking cows. That is the reality of what it's like to be a farmer, and whatever mechanism I could use to help my father and people like him, I wanted to use.

"There is an incredible capability of the tools of agricultural biotechnology to help farmers. What amazes me is that the people who say we need to look out for our environment and for the health of consumers are the very people who are trying to force us into a position in which we have to rely on old-fashioned, less effective, and more environmentally damaging techniques."

For devotees of biotechnology, the monarch butterfly disclosures marked the first public-relations crisis. The defense they mounted capitalized on an emerging truth about the deployment of genetically modified seeds. There indeed may be costs to planting the GMOs. But in the calculations a society makes about a new technology, these costs—environmental or otherwise—must be stacked up against the benefits. The biotech defenders will quickly turn the tables on the critics, as McGloughlin showed in questioning the safety of organic farming.

"Think about it," she said. "If you don't have Bt pollen, you'll be using chemical insecticides that kill monarchs and everything else, too. Given everything else that is going to hit them—Mack trucks and disappearing ecological niches—the problem from the pollen is tiny. I look at it as a matter of a cost-benefit analysis to see if it is worthwhile. In this case, the benefits so far outweigh the costs—and there may be a cost— that it's not even an issue.

"You have to use techniques that are going to allow you to control things like fungal toxins, which can cause liquefaction of our brains and cancer in our children. Using organic systems is not sufficient. One of the mainstays of organic agriculture is natural fertilizer. But if it's not properly composted, you get e-coli. I tell the story about the health inspector tracking down an outbreak of e-coli [bacteria]. He finally traced it back to a farmworker who had relieved himself in the field.

"I think it's terribly disingenuous for anyone to say that, by using purely organic mechanisms, you can create sufficient food at sufficient volume that it can be priced at what people can afford. That's elitist. Some of us are in the wonderful position where we can totally live off of organic food. But you cannot feed the world. You cannot even feed the inner cities."

Often, McGloughlin finds herself defending perceived threats to people as well as to the environment. The spread of genetic materials from modified plants to weedy relatives is among the environmental threats whose consequences are unknown. But concerns about allergic reactions could prove to be a bigger consumer threat. McGloughlin insists that there's little reason to worry now—and even less risk in the future.

"We don't have to worry about escaping genes. First of all, if that happens, you need about twelve different components, and we have about three hundred chemistries to deal with it," she counters to deflate the GMOs-on-the-rampage argument.

"Our plants, we treat them like princes. We mollycoddle them, give them the best of everything. If we let them go wild, they will throw off these [engineered] traits. They will throw off everything that doesn't give them a competitive advantage. Because it really is a jungle out there.

"Soon, we will be using chloroplast transformation [in genetic engineering]. It will completely control gene flow. You will get a higher level of expression of the gene. And the chloroplasts are transferred purely on the maternal line. They stay in the mother's cell."

Turning her arguments to GMOs' effects on people, McGloughlin argues that precautions are sufficient and in place. "As far as allergenics," she says, "the number of tests done to look for them is enormous. There are so many checks and balances. Think about it from a company's perspective. It would be a terribly dumb business move to harm consumers. This notion that you are putting products out there willy-nilly is amazing to me. They talk about rats getting stomach ulcers from [engineered] tomatoes. If you were eating tons of tomatoes full of acid, you would get stomach ulcers, too. Things are often being put out of context. It's often difficult for a nonexpert to look at this."

McGloughlin sees a future in which people rely on genetic engineering not just for the efficiency it can bring to farming but also for its benefits to human health. It is this prospect of more healthful food—improvements that may not arrive until late in the first decade of the new century—that the biotech industry embraces as a shield against attacks on its technology.

"With these products," she says, "consumers can take a more healthful approach in their lives, rather than running to the doctor. They can be proactive rather than retroactive. In the laboratory, they can modify proteins and carbohydrates at the macro level, introduce vitamins, minerals, antioxidants; put antioxidants in broccoli, for instance. They can insert anticancer agents."

McGloughlin intends to remain in the vanguard of the pro-GMO forces. She is frustrated that other scientists haven't joined her in what promises to be a long public battle.

"I think more scientists need to be made to feel that they need to be out there. Too many of them think the opposition will just go away. I think that for a while, there will continue to be a debate over trade issues and genuine concerns about the technology. I think we are in an extended transition period. But if you look at what is happening, farmers are adopting it faster than any other technology in history."

MARGARET: ON THE STAGE, BIOTECH "WANTS TO BE THE STAR"

In the other corner is Margaret Mellon. Mellon, born in 1945, grew up in Iowa, the perennial champion of corn production in America's belt of muscled farm states, looking across the Mississippi River into Illinois, the number two state. She admittedly knew little about agriculture as she headed eastward on a peculiar educational odyssey first to Purdue University and then to the University of Virginia, from which she emerged an unusual genetic cross: a lawyer with a doctorate in microbiology.

In the mid-1980s, when policy makers started playing catch-up with scientists, she was an attorney with a well-known Washington firm, Beveridge & Diamond. You might have been able to count on one hand back then the Washington lawyers well tuned to the complexities of biotech confronting Congress and the White House. What she was hearing about the marriage of genetic engineering and farming appealed to her, both as a scientific discovery and as a specialty in a city that demands specialization.

"I'll bet if I was writing a diary then, I would have seen myself as somewhere in the middle of the debate, trying to explain genetic engineering to people and telling them that it was nothing to be afraid of. Kind of a midwife to something that might have some problems but that, by and large, would be great," she said.

Mellon later joined the Washington-based Union of Concerned Scientists, where she grew skeptical about the technology she had welcomed. Since then, Mellon, who is also a law professor at the University of Vermont, has become one of the most knowledgeable critics of genetically modified farming. She believes that genetic engineering may be a useful tool, but she sees no evidence that it is living up to its promise, let alone becoming a panacea.

I also asked Mellon to tell me her story and how her thinking evolved.

"My father worked in a manufacturing company: fiberglass. It was a corn desert in Iowa, but the Mississippi River was flowing at our backs. I knew that a lot of my friends lived on farms and we were proud of our farms. But I didn't know anything about agriculture. What I brought with me from Iowa was a feeling that agriculture policy was a given, that of course everybody grew corn.

"Early on, I was pretty high on the technology. I was the perfect molecular biologist. I was so impressed by the technological achievement of being able to move genes back and forth that I had a hard time understanding that there might be limitations. I was predisposed to see science as a set of steps, and this was the next step.

"Fifteen years later, there is a convergence of different streams: of trade and the environment and the horrendous developments in agriculture; the fear of monopoly and the rise of corporations. There is this convergence, a layering-on, that gives this issue gravity and density that it never had before."

Mellon admittedly knows of no proved health threats from GMOs. But the uncertainties remain, she argues, because of all the complex interactions in genetic engineering. In biotechnology's brief but stormy history, she asserts, many of the initial assumptions have already changed.

"In each case, you have to think of what genes you've added, what pieces of DNA you've added, the genetic background of what you've added—be it a crop, a fish, or a fowl—and then try to think through those new combinations of traits in the new background and what might go wrong. You have to basically ask it product by product. It's an important question and a very difficult question to answer.

"There's been a trajectory to this debate," she argues. "At the beginning, who knew where it was going to go? There were all of these reassuring scientific statements that were made in the late eighties. That nature has tried all the combinations of genes and all the ones not represented by today's organisms are failures, so we can't try anything that nature hasn't already tried and rejected. There was the idea that genes really won't flow, and that there won't be horizontal gene transfer. The whole scientific argument has changed. Now they say, 'well, horizontal gene transfer is taking place all the time and you don't need to worry about it anymore.' The scientific underpinning of this debate is changing, and it is changing to be less and less reassuring."

To Mellon, the monarch research isn't close to being debunked. The results at Cornell, she says, provide a perfect example of the quirkiness of genetic engineering. Underlying this debate, she adds, is another risk that lands heavily on the side of the ledger that measures the costs of biotechnology: the looming loss of Bt, which she describes as a "gift of nature," thanks to the zeal of the genetic engineers.

"The whole issue," she says, "is the nontarget relatives of the targets. If you put Bt into corn, you have to ask yourself, in addition to getting the European corn borer, is the Bt also going to get the relatives of the European corn borer? In this case, that is all other Lepidoptera, including the monarch, the black swallowtail and the luna moth.

"Nobody wanted to put the Bt into the pollen. It's just that they can't target the genes well enough to put them where they want them. That means you've got to think through not just what would eat the pollen but this whole extra layer of interaction that results from the pollen being wind-blown away from the corn and falling on a plant. You just don't think like that when you're talking in conventional breeding terms.

"There are problems we think are associated with these genetically engineered crops. You can talk about losing Bt. Clearly, genetic engineering allows you to put Bt in the environment in a manner that produces enormous pressure on pests to develop resistance. And it's certain that some classes of insect pests will develop resistance. It's just a matter of time.

"It's a big problem because it isn't just that the genetic engineers lose this important protein, which is turning out to be even more valuable than people thought it would be. It's the other people, who didn't get even short-term benefits from the use of Bt, who will be deprived of it. People in conventional agriculture, under pressure from the Food Quality Protection Act, are trying to use more and more Bt. The organic farmers look to Bt. This is a really valuable gift of nature that, from the genetic engineers' point of view, if they get five years of profits out of it and then lose it, they probably don't care. It's part of the hubris of biotechnology that there will be these other genes coming. The other genes have been promised since 1988. I don't know of one that is near commercialization. I know of a lot of them that have been tried, but they don't work; they're not as good as Bt."

Unlike McGloughlin, Mellon does not dismiss the risk of allergens. And she sees a host of environmental concerns that range from damage to fish stocks to the uncertain potential of "jumping genes."

Of potential effects on humans, she says, "Allergens are in some ways unique to genetically engineered plants because no other technology can bring in as many proteins, and therefore allergens from unrelated organisms can pose a threat of hidden allergens. Experience tells you a lot if you stay in the confines of traditional breeding. If you're not allergic to a tomato, you're not likely to be allergic to new varieties of conventionally bred tomatoes. But not being allergic to tomatoes tells you nothing about whether you're likely to be affected by a tomato with a fish gene in it."

Mellon is just as wary of the environmental consequences of GMOs. Take, she says, the example of salmon.

"They now have been able to literally screw up salmon metabolism in such a way that the salmon grow very big, very fast. By the end of their lifetime, they have not achieved any larger dimensions than normal salmon. But they grow very, very fast because their cycles of hormone production have been disrupted.

"That could be an enormous problem from an ecological standpoint because these rapidly growing salmon are going to be much more attractive to potential mates in the wild, should these fish escape. Although they are sexually more attractive, from a fitness and environmental standpoint, they are not going to survive. That means you are going to drain off the reproductive potential in the normal population with basically an unfit organism."

As this opponent sees it, that's only the beginning of a long list of problems.

"We need to worry about jumping genes," she insists. "We know that these genes are going to move through the crop into wild organisms. So we've got all kinds of gene-flow problems. It depends on where those genes are going to end up and what they are doing. Salt tolerance is a good example. If you have salt tolerance move, say, from a salt-tolerant rice plant to a rice relative that's a weed, and that gene can now move to nearby coastal wetlands, you can destroy coastal wetlands with invading plant species.

"We now know that a lot of genomes are made up of a lot of junk DNA; some of that junk DNA appears to be remnants of mobile pieces of DNA that can move all around the genome and occasionally from one genome to another, crossing species barriers. We may not want these lines crossed more often than they now are."

Many of biotech's promises have not been kept, she argues, and may never be realized because of the complexity of the technology.

"There is no way with current technology that people can even think about moving twenty or forty interacting genes into plants to solve all the agronomic problems," she says. "What about salt tolerance? And what about drought tolerance and increased intrinsic yield? Those are three things that agriculture could really use, particularly in the developing world or anywhere else facing stressed conditions. People talk about it all the time. Is there any reason to think that genetic engineering is a better way to get there than by traditional breeding?"

"I picture people sitting around with a confetti of genes, and out of the fifty thousand, finding the eight or nine that probably have something to do with drought, have something to do with yield, have something to do with salt tolerance. Then fish those out, figure out how they work, and put them back into a plant, adjusting their relative rates of expression and interconnection so that they mimic what you would find in drought-tolerant plants.

"That's not the fast track. Maybe twenty years from now it will become a fast track. Maybe people by then will have a great knowledge of what the genes do, how they interact, and computers will be able to select the important genes. But I'm not even sure that's the case. I think the idea that genetic engineering is a faster track to the important agronomic end points doesn't ring true. It doesn't ring any more true than for somebody to be promising nitrogen fixation. And yet people still talk about it."

What troubles her the most, Mellon says, is the attention to a technology that has provided so little. It's part of our romance with technology, our "yearning for silver bullets," as she puts it.

"There is a problem that when you're dealing with a technology like genetic engineering, the social, political, and economic dynamic that drives this technology forward has enormous opportunity costs. It basically drives everything else off of the stage. It turns all of these important questions into, 'what do we do with biotechnology?' rather than 'how do we help feed the world?' Or how do we improve the nutritional status in the United States? Or how do we improve the economic status of the farmer? All of those questions get turned around to, 'Can you use it to solve this problem, or that one? Can you put enough money into it so that its promise will come out?'

"The big ideas are all being played out," she says. "The notion that it's inevitable. The notion that somehow its glitziness means it's preferable and that it will perform better than what we've had before. I see biotechnology on a stage. It's the loudest. It wants to be the star. It's knocking other players off of the stage. If it were only modest, and would let other people play their part and it kind of stood quietly off to the side, and said 'here are a few places where we think we might help,' you might be able to welcome it.

"But as long as it's elbowing everybody off the stage, demanding the spotlight only for itself and sucking up all the resources, and leaving everyone else starving for air, you have to be concerned about it even if you think there might be some good uses somewhere down the path."

PLANTINGS THREE

The day my soybeans popped out of the ground, Percy
Schmeiser went on trial in Saskatoon for seed piracy. Schmeiser
is a Canadian farmer, past mayor of the town of Bruno, and a
former assemblyman in the province of Saskatchewan.

I don't know if Schmeiser is guilty of allegations brought
by Monsanto. I do know that he's a remarkable fellow. I can't
name any other farmers who have climbed Mount Kilimanjaro
and attempted Mount Everest three times, making it almost to
the summit.

In Saskatchewan, roughly half of the canola farmers were
drawn to Roundup Ready seeds by the time the new century
arrived. They plant their seeds, spray on the Roundup, and
the quack grass and the rest of the troubling weeds die while
their plants flourish. Schmeiser, who is pushing seventy and no
longer challenging the world's tallest peaks, swears he's no pirate. He says he didn't need the modified seeds, that his conventional seeds had adapted to the soils out in the Muskiki hills
and yielded bountifully.

Monsanto disputes this. In his opening statement, the company's lawyer argued that 90 percent of Schmeiser's canola crop
was one of its genetically engineered varieties. "It's just too
much," he said.

On the United States side of the border, Monsanto uses
Pinkerton agents. In Canada, court documents alleged that
Monsanto had hired retired Royal Canadian Mounted Police
to track down farmers suspected of planting engineered seeds
without paying technology fees. Schmeiser contends that investigators sneaked onto his land and that Monsanto representatives had showed up at the Humboldt Flour Mill, where he'd
sent his seeds for cleaning, asking for samples.

Schmeiser's lawyer told the Federal Court that the canola

in question had suspicious DNA because it had become crossed from engineered pollen blowing in the wind. Bringing modified seeds to Canada was like letting "a genie out of a bottle," the lawyer said. "It was something that was unleashed into the environment and cannot be controlled."

Considering what I'm growing, the testimony makes me nervous on more than one score. From what I have read, it seems that Schmeiser has become a cause célèbre among anti-GMO activists and farmers feeling oppressed. Old Percy even has his own web site. It occurs to me that even if Monsanto wins, they lose.

Part Two

ON THE PHARM

6

IN ILLINOIS, AN APOSTLE OF MODIFIED FOOD

*Insects of the day spend their brief existence in reiterated coition, lured
by the smell of the inferiorly pulchritudinous extendified pudental
verve in the dorsal region.*
—James Joyce, *Ulysses*

FARMER TIM SEIFERT shakes his bony index finger at the environmentalist
whose shoulders are flush with the back of an elevator in Crystal City, Virginia. Each has addressed a U.S. Environmental Protection Agency advisory
panel studying the perils of growing corn that is genetically engineered to
produce its own insecticides. The hearing went smoothly, the exit less so.

The GMO critic had argued that heavy plantings of Bt crops will hasten
the evolutionary cycle of pests and thereby allow them to develop resistance to the *Bacillus thuringiensis* bacteria. That would be bad for everybody
except the bugs. Gardeners and organic growers—not just the genetic
engineers who have heisted Bt's magic for their high-tech crops—have
depended on the naturally occurring bacteria for seventy years.

Tim, by contrast, sees no threat from Bt plant pesticides as long as
farmers use refuges—acreage planted in conventional crops—to give the
pests a place to feed and reproduce, helping to forestall resistance. Tim
had come to this Washington suburb, on a ticket bought by Monsanto,
to talk about the care deployed by state-of-the-art American farmers in
the Genetic Era. I thought that he had spoken persuasively.

Tim is an innovator from the git-go, not just with genetically modified
seeds. He and his father, Ed, invent farm machinery, such as their hydraulic

contraption that spreads chaff evenly in fields for their no-till, antierosion farming. Another implement, their stalk retriever, was a hit through the Midwest until the Occupational Health and Safety Administration suggested it might be, well, risky. Tim experiments in the rich, dark soils of central Illinois with dozens of corn and soybean seeds. He wants to see what sprouts strong and yields the heaviest, so, come next spring, he can lay down his smartest wager, like bettors at the ten-dollar blackjack table on one of the riverboat casinos in St. Louis, ninety miles down old Route 66.

Tim Seifert was an early convert to genetically modified crops, and he'll plant any gene-altered new seed that will shake out of a bag. He conducts his own personal field trials not just with corn and beans but also with engineered sugar beets, canola, and cotton, crops almost never grown in his neck of the prairie. The artsy-crafters around Auburn and Chatham love to get their hands on those exotic tufts of cotton.

Tim Seifert is a poster boy for farming. He's in his early forties, handsome and six-foot-four; he was a long-legged sprinter in his sports days. He's a Ford-loving Farm Bureau member, a trustee for Trinity Lutheran Church, a member of the school board, and a basketball coach. He was chosen in 2000 as one of thirty members of the Illinois Agriculture Leadership Foundation. He even finds the time to volunteer as an ambulance driver. He's just the sort of fellow you'd want talking up your jazzed-up new seeds.

Truth be told, Tim doesn't especially enjoy being around a crowd in suits. If he has a shortcoming, it's a tendency to get a tad worked up. Once I referred to him in a magazine article as "somewhat excitable," and when a neighbor brought him my article, his reaction was: "Well, that sounds like me." In the elevator with the environmental advocate after that EPA meeting, he was living up to the reputation.

"You go home from your $65,000 job and you cook your steak and you go out in your garden and pick a tomato. I'm trying to do what's right for the world. You're trying to do what's right for you," Tim says to him.

"I guess I'd better get off this elevator," the startled man replies.

"I guess you'd better."

I could see that Tim Seifert was somebody to know.

Farmer and biotech devotee Tim Seifert watches over young corn plants on his central Illinois farm. He believes that genetically modified seeds "have made a lot of bad farmers good farmers."

CENTRAL ILLINOIS: LOURDES OF BIOAGRONOMY

I'm from central Illinois, too, and sometimes I think that the bubbly, nutrient-loaded soil there sprouts farmers like Tim Seifert, people whose shoulders are squared with certitude and who keep ass-whuppin' sticks in their back pockets. Tim told me once, "My reputation in the community is that I don't stand in line for a ticket. I write my own."

I felt like I knew Tim even before we'd shaken hands after the EPA hearing, and when I introduced myself, I'll be darned if we didn't know some of the same people in the rural reaches of Springfield, which is fifteen miles from the Seifert home on Panther Creek Road. Over the next two years, as I watched the European GMO debate sizzle, then flame, I telephoned Tim every few months to see what he thought. The roiled European markets complicated his decisions about what to plant. But I could tell that he was just as troubled by the repudiation of American agriculture. I'd ask him, how many of your soybeans will be genetically engineered this season?

"Every inch," he'd say. "When are you coming for a visit?"

Tim's farm became a tour stop for foreign delegations in trips put together by Monsanto. They'd fly farmers over from the European front lines of the GMO wars, and from Brazil, where wealthy landowners in the desolate Cerrado were hankering to keep up with their Argentinean neighbors and get those Roundup Ready soybeans in the ground. They'd come to Middle America from Japan, a big importer of American grain and a country coming down with a case of the GMO jitters.

When promoters wanted to show the miracle of genetic technology at work, Seifert's was the place to take them. Tim was affable, successful, and unabashedly keen on the cost-saving, sweat-reducing miracle of bio-technology. Monsanto didn't have to give him a nickel to show off their science in his fields.

Tim had been enthusiastic from the start about the new seeds. Zealous, some would say, among them a visiting Japanese soybean crusher who almost ended up like the fellow on the elevator. Crushers process soybeans into oils, meal, and any of the host of products that support processed foods. At Tim's farm, one of the Asian visitors voiced reservation: "Would you be willing to start sending us non-GMO, identity-preserved soybeans?" he asked.

Tim cast his gaze toward his nine hundred acres of soybeans, planted every inch that year in genetically engineered seeds. He didn't answer the question until he and his visitors sat down at a steak house in Springfield. Then he leaned toward the questioner.

"You people tell me what you're afraid of," he said. "Go on, tell me. You, of all people, are saying you're afraid of technology. If we had been afraid of your technology, we'd still be using sundials."

Tim cooled down before he finished answering. "Yeah, I'll go non-GMO, but you'll have to pay me a premium to do it."

—PROS- (Farmers)

LAND OF CORN AND BEANS

Tim Seifert is wearing a green polo shirt with a logo I've seen before: Monsanto—Food, Health, Hope; next to those words is a sprout of grain.

Before I head west toward Seifert's from I-55, which locals still call by its old name, Route 66, I see a white billboard in a field proclaiming, ROUNDUP READY TEST PLOT. Monsanto may be "Mutanto" in Europe, but the ridicule hadn't reached Middle America. Farmers proclaim their brand allegiance like teens displaying sneakers, and Monsanto stuff remains in demand.

Tim and Roxy Seifert have two children, Jordan, ten, and Weston, five. Outside their white-clapboard farmhouse, an American flag fills with a fresh westerly breeze. I intend to ask Tim why so many American farmers are converting vast swaths of ground into factories for genetically modified food. I'd like to know why they got in cahoots with the life-science behemoths in the first place, and why they sign seed contracts with terms that can land them in jail. If he doesn't get too worked up, I might ask how it feels to have much of the world telling you to stick your funny grain back up your combine.

First, there's trouble. A freakish wind has zagged through his fields, ravaging tiny corn plants. We're headed north to Springfield in Tim's F-150 when he spots the plants drooping. "Stripped down pretty tight," is his way of seeing them. He gets on his Nokia and leaves a message for an insurance adjuster. There may be losses, and he is worried.

"We watch these crops grow. We caress them. When something hits— like hail, which is the worst—we feel it. It's like something happening to your kids," he says.

It's just after eight o'clock in the morning when we enter the federal Farm Service Agency office off Stevenson Parkway in Springfield for the annual ritual of certifying crops for enrollment in federal farm programs. On aerial maps spread over the counter, I can see what Tim has planted, and where. This season, Tim is farming twenty-four hundred acres and some change, half of it in soybeans and half in corn.

Fifty-fifty soybeans and corn is a common split in these parts, and the crops typically are rotated from year to year. Farmers in central Illinois grow almost nothing else, even though prices for each reached generational lows at the end of the century and continued heading south. Few farmers plant wheat because they no longer need the hay for livestock. Like Tim, most farmers don't bother with animals—unless they choose to turn over their land to corporate hog producers and make a living by tending the company swine. They have to be desperate to do so, or close to it, and resilient enough to fend off neighbors and townsfolk who are likely to organize to forestall the holy-hell stink. Here in Sangamon County, you wouldn't want animals clomping up four-thousand-dollar-an-acre ground that is unfailingly fertile for corn and beans.

In the truck headed home, I probably didn't need to ask how many of his freshly planted beans are genetically engineered this spring. But I do.

"Every inch," I hear.

Tim is so good at what he does that he grows soybeans for seed companies. The seeds have been engineered for herbicide tolerance, and they pass on this trait to the next generation. Any brand of Monsanto-invented glyphosate will run harmlessly off these plants like water off a duck's back, even as the chemical chokes yellow foxtail, velvet leaf, and any other vegetation farmers don't desire in their dirt. By spraying a solution enriched with a quart or so glyphosate per acre, you kill the competition for nutrients and give your plants room to spread. It's a system: Plant the seed and then spray the proprietary chemical for which it's genetically engineered.

Farmers will tell you they do it for profit, for ease in farming. They do it for vanity, too. You keep a "clean" field year-round, naked of anything except what you sow, and your land looks good and rents for more.

"Roundup Ready has made a lot of bad farmers good farmers," Tim tells me.

Tim owns 180 acres and leases another 2300 from people too old to farm or from the heirs of dead farmers. He is obsessed with finding the best seeds to plant, and in a padlocked old truck trailer he keeps dozens of varieties. It's like a wine cellar for seeds. In each fifty-pound sack of soybeans are enough seeds to plant between 2.5 and 2.8 acres. A bag costs about eighteen dollars and onto that you tack a six dollar "technology fee" for using a company's genetic engineering system. Farmers sign contracts when they buy the seeds, and if they forget what they've agreed to, the label will remind them.

One of the labels on Tim's seeds reads: "THIS SEED IS ACQUIRED UNDER AN AGREEMENT . . . solely to produce a single commercial crop in one and only one season. This license does not extend to the seed from such crop or the progeny thereof for propagation and seed multiplication. The use of such seed . . . is strictly prohibited. Resale of this seed or supply of seed to anyone for planting is strictly prohibited."

This spring, Tim has planted test plots with sixty-three different varieties of soybeans, thirty-three of them genetically engineered. His industriousness amazes me. He's applying seven different formulations of glyphosate herbicide to see what works the best. The cab of his seventy-five-thousand-dollar Case, one of three tractors, has more computer power than an early spacecraft. There's a Global Positioning System unit receiving from satellites, and when he powers up his built-in laptop, he can retrieve a schematic of any of the forty-six fields he farms. As he moves through the field, his tractor a dot on the screen, the computer displays variations in soil types. The laptop shows the bushels-per-acre yield in past years and the date the crops were planted. It tells him what, when, and where to plant. If it's especially rich soil and he's planting corn, he might sow thirty-five thousand seeds per acre. If it's not his best dirt, he may dial down to twenty-five thousand.

In his field truck, another computer linked in the Global Positioning System enables him to spray a precise formulation of herbicide. Likewise, he can apply as much nitrogen fertilizer as the soil needs, usually about 150 pounds per acre every season.

"People think we're just out there blowin' this stuff out," Tim says, shaking his head.

"I'm independent, you know what I mean?" Tim says, when I ask him if he enjoys being a farmer. "I'm independent; I come and go as I please. But I've still got a ton of people to answer to. I've got eighteen landlords to answer to, a banker to answer to, a family to answer to. I have to be self-disciplined; I've got to go out and get it done. When the ground is wet, I have to sit and wait to plant or I won't get the best crop. And when it comes time to harvest, I have to be disciplined. If I go in too early, I take a yield loss.

"I have to be a scientist, an agronomist, a weatherman, and a mechanic. And now I have to be a computer expert. And, of course, I have to be a money manager."

TROUBLE WITH ENVIROS

Nothing lights Tim's fuse like an environmentalist. The first time I met him, I quoted him in the newspaper as saying, "They tell us that in the next century, farmers will have to grow six hundred bushels of corn per acre to feed the world or we'll have to tear up the rain forests for planting. Then what will the environmentalists say?"

I'd heard other farmers say as much, especially American farmers plowing hefty acreage. They want credit for growing food, for feeding people, and I think they ought to get more of it. But seldom do they hear a kind word from environmental advocates, who tend to paint farmers as chemically addicted destroyers of the earth.

On water pollution, wetlands, and virtually every land-use issue that comes up, the American Farm Bureau and its allies stand far apart from the Sierra Club and other environmental advocates. In reporting on these issues over the years, I watched an uncommonly wide gulf open between farmers and environmentalists. In the 1990s, yet another divide opened: damage from farm fertilizer runoff, calculated in a seven-thousand-square-mile "Dead Zone" of oxygen-depleted water in the Gulf of Mexico where the Mississippi River dumps in. Farm-nutrient pollution also chokes rivers, bays, and lakes with algae blooms. It is a vexing problem with no ready solution and little cooperation. In Congress, adversaries almost always work together. Farm and environmental lobbyists seldom speak.

Now they are being driven even farther apart as more American farmers embrace genetic engineering and more environmental groups join the opposition. When I ask Tim about the widening gulf, he says that it's been on his mind.

"I think about it every day. The question I have is, why are they taking a negative view? Why does everybody have to have such a negative attitude? Why do they have to think that it's a bad product? Why would Monsanto and Novartis want to tear up the environment? I find it hard to believe that these people think these companies are trying to kill us off.

"I have to put my confidence in somebody. We have to have some kind of faith in something. This day and age, a farmer needs to align himself with somebody. If you don't, it's you against the world. Everybody essentially is tied in with somebody. You can be a survivor, but it's getting harder every day. Do you know what I mean?

"I tell Monsanto, 'you gotta be straight with me, I'm the one that's out front on this. I'm in the coffee shops. If there might be a problem, tell me now.' The way I look at it, if Monsanto wanted to kill off the world, they would pick something better than corn and beans."

As if on cue, a Monsanto sales rep in a black Dodge Durango drives up to the edge of a cornfield where Tim and I are talking.

"It's people like him we need to speak up more," says the Monsanto man, Craig Riley.

HARD CHOICES

The tender leaves of foot-high, gene-altered corn twitch in the breeze like unruly toddlers. On a clear day like this one, you can see far across the Midwestern prairie until, it seems, the earth starts its curve. What I see are annuals in monoculture as opposed to perennials in the polyculture that this prairie once supported. All the way to the horizon, fields are greening with fresh crops, many of them genetically modified.

The poet Carl Sandburg said that the denizens of the flatlands achieve wisdom because they can see so far in the distance. I can't say I know what will happen as far as acceptance of genetically modified food in the world. But standing in this field of genetically modified corn, I am able to see that the future holds nervous times for farmers like Tim Seifert.

From talking to his neighbors, Tim estimates that 95 percent of this year's soybean crop in central Illinois is genetically engineered. It's reassuring to them that the Europeans have not rescinded their acceptance of foods with American soy ingredients. Corn is a different story, and Tim's recent decision shows the limits of a farmer's allegiance to a technology or a company. Tim insists that the new stuff works great; last year, his genetically engineered corn produced 194 bushels per acre, compared to 178 bushels from conventional seeds. In an average year, genetically modified seeds will produce the same or more as regular seed, Tim says—an assertion not always supported by research. "In a dead-even year, it's a break-even deal," is his way of putting it. But in what he calls "stress-related years"—seasons of uncommon pest infestation, drought, or otherwise freakish conditions—the modified seeds are more profitable because they protect the plant.

PRO for farmer

There are other calculations involved—and other stress than from Mother Nature. Bt corn is engineered primarily to protect against the European corn borer. But Tim hasn't seen so many of the pestiferous worms of late, and he must calculate the extra cost of the modified seeds and wonder if the warm, wet winter will mean more pests. He's thinking about his buyers, among them tortilla-makers and livestock-feed producers. Tim Seifert also is thinking a lot about the anti-GMO tide across the Atlantic. With soybeans, the Seifert farm remains "every inch" planted in modified seeds. But this season, he has scaled back his planting of modified corn. In the past, Tim planted as much as 40 percent of his corn in seeds engineered for insect resistance. In other words, rather than planting five hundred acres with the new technology as he did three years ago, he's down to about two hundred acres.

When I ask him why, he speaks about "this crap in Europe in play. It seems like everybody runs on the coattails of the European market, do you know what I mean?" he says.

In making his planting decisions, Tim considered the prospect of being paid extra this year for growing nonengineered grain. Some of his buyers, he speculates, might worry that they can't sell their corn in Europe, even for animal food. "Wherever that product ends up, we have to be careful."

But Tim insists that he is not abandoning the technology of genetic engineering. "This is not about economics, about an easier way of life, or whatever. It's about conservation, about health, about hope for the future. Today, we have grain bins full of corn. But it might not always be like that. We might have that drought they keep talking about. We might have bugs eating up the crops.

"This GMO thing is right, and farmers know it's right, and we're doing our best to tell the world that they're missing the boat. But we also have to survive, and from an economic standpoint, we have to have all of our bases covered," he says.

Occasionally, a sliver of uncertainty creeps in, even in a guy wearing a Monsanto shirt. "Fifty years from now," Tim muses in his pickup, "we'll be sitting back in our rocking chairs, and we'll be saying 'we were right.' Or they'll be saying they were right."

HARVEST

When I telephoned Tim five months later, he had just begun harvesting, and the news was mixed. His crop was a bin-buster. Not only had down-state Illinois received plenty of rain, it rained at the right time. Once more, he said, his two hundred acres of Bt corn outperformed his conventional varieties, producing 240 bushels per acre.

The bad news was that the price for all that corn had plummeted to an abysmal $1.50 a bushel; three years earlier, it had been $5.25. "We've got more corn than we know what to do with," he said.

What's more, the European corn borer was back. Tim spotted the tiny brown worm burrowing into stalks, leaving them brittle and vulnerable. But in some of his fields, he had Bt corn out there doing battle. And he declared that next year, he'll be planting a full measure of it—despite enduring another round of bad publicity about threats to the monarch butterfly.

I asked him how the public-relations effort fared, and he sounded weary. "I'm really burnt out on it all. It's a pain in the ass. The Sierra Club and the Greenpeacers think we're out here trying to tear up the universe. They're not looking down the road. They're satisfying their own hunger all the time. But it's my responsibility to feed the world."

It's a "no-brainer," he said with regard to his choice of genetically engineered crops, before inviting me back to the farm for the remainder of the harvest. "You can ride the combine," he said.

7

THE TERMINATOR

I'll be back.
—Arnold Schwarzenegger, *Terminator,* 1984

DELTA AND PINE Land Company is situated in Scott, Mississippi, on what once was the biggest cotton plantation in the world. It covered nearly sixty square miles, and in its hospital, plantation doctors tended to the health of sharecroppers. There was nothing like it in the world. In the 1930s, a socialist group praised the company for handling its contracts with croppers fairly and dispensing sound advice to them. On the whole of the Mississippi Delta, a skinny diamond of land that stretches from just beneath Memphis down to Vicksburg, nobody could grow the "white gold" like Delta and Pine Land. In 1936, when cotton patches in the American South averaged 200 pounds per acre, D&PL's fields yielded 638.

It was a powerful and enduring company and, like many early-twentieth-century corporations, Delta and Pine Land knew how to manipulate the system. Indeed, its name, which suggests something other than a cotton grower, sprang from a need to circumvent the law. The British investors who started buying up delta land around the turn of the twentieth century discovered that Mississippi law prevented agricultural holdings of over ten thousand acres—unless they were chartered in the 1880s. So they found themselves a grandfathered company, a land-speculation outfit called Delta and Pine Land, and three companies merged under that name.

In the 1930s, company president and erstwhile gubernatorial candidate Oscar Goodbar Johnston grew rich growing cotton and investing in cotton futures. He was a promoter as well, establishing the National Cotton Council to push the needs of his industry. Johnston also discovered that he could make good money from cotton by not growing it; in 1933, a time when many American farmers were going under, he received $115,000 from a government crop-reduction program.

Later, the company was hacked into pieces by its British owners, and the Delta and Pine Land that emerged in the second half of the century was primarily a cotton-seed and research company. It grew powerful in its own right, corralling nearly three-quarters of the cotton-seed market in the United States by the 1990s.

In 1996, through a collaborative agreement with Monsanto, Delta and Pine Land became the first company in the world to sell genetically engineered cotton, an insect-resistant variety called NuCOTN. The company would grow even closer to Monsanto, while nurturing its working relationship with the government. From this cooperation emerged a discovery that would change the course of the global biotechnology debate: the Technology Protection System, known to the world as the "Terminator."

Genetic engineering conferred on plants an attribute heretofore unknown: It rendered their seeds sterile. In other words, the wheat, rice—or, theoretically, any plant in the world imbued with the system—would be the last of its line. A farmer collecting its seed for later planting would be sweating for naught.

The newly patented system offered hope for managing the undesirable consequences of genetically modified crops. Farmers wouldn't care to see the newly engineered trait of herbicide-tolerance spread to the weeds around their fields. Turning on the Terminator could limit the escape of a trans-gene into the wild, thereby preventing the potential for outcrossing, emergent superweeds, and other unwanted alterations to the environment.

Such threats were not much talked about, at least by genetic engineers, until time—and need—came to defend their Terminator.

But the capacity to prevent genetic pollution was not why the technology was created. Nor was it a feature that would go far in assuaging skeptics. The Technology Protection System was invented to protect the intellectual property of the companies engineering the new agronomic traits, principally herbicide tolerance and insect resistance. Farmers would be forced to buy seeds at each planting time, and this prospect frightened and angered seed-saving farmers around the world.

Terminator was a term coined not by Delta and Pine Land but by its detested detractors. But it was such a devastatingly apt shorthand that it came into wide use not just by critics but by companies themselves; even by then–Agriculture Secretary Dan Glickman.

The technology developed by government scientist Melvin Oliver at the Agriculture Research Service labs near Lubbock, Texas, along with Delta and Pine Land's Donald Keim and other scientists, is not a single, omnipotent gene, as is widely believed, but a three-gene construct. The first gene, a so-called promoter, generates a toxin that obstructs the gene's capacity to germinate, while leaving the rest of the plant normal in its physiology.

A second new gene enables the control system to operate by producing an enzyme that breaks down a wall of genetic material in cells. Finally, a third gene, carrying a chemical (typically the antibiotic tetracycline) makes proteins that block production of the second gene's enzyme. Until the plant matures, the sterilizing toxins do not kick in. The vegetable or fruit is otherwise normal.

It is a complex and elegant system to subvert nature, and a technology for which Delta and Pine Land, which co-owned the patent along with the Agriculture Department, had high hopes. Not only would it create a business system that catches up with science, it would install in the field a policing system to quash genetic pollution. It would, therefore, quell an emerging worry about installed genes: their escape through the country-side into unsuspecting plants. Of course, it couldn't escape itself. Delta and Pine Land's Harry B. Collins tried to put the technology into per-spective for me.

"It's like the old playground joke. It's a scientific fact that if your par-ents never had children, there's a high probability that you won't either," he said.

FLUSH OF DISCOVERY

I visited the old Mississippi plantation in April 1998, five weeks after patent No. 5,723,765, with the vague title "Control of Plant Gene Expression," had been issued to Delta and Pine Land and to "the United States of America, as represented by the secretary of agriculture." The patent had provoked little coverage in the United States; my story would be the first beyond bare mention. I had hoped on this spring afternoon to be able to ride one of the golf carts around the maze of buildings, but since company president Murray Robinson's office was situated near the entrance, I didn't get my ride.

Robinson, an erudite man near retirement, was just as sold on genetic-engineering technologies as on his company's invention to protect them. By this time, modified cotton seed already was producing 40 percent of the nation's crop. As he accurately characterized it, Bt cotton had created a revolution in his industry. "All cotton farms have farmhands that didn't graduate from rocket science school," Robinson remarked by way of describing how gene-altered cotton simplifies farming, taking the guesswork out of which seeds to plant and which insecticides to spray.

TPS, the industry's shorthand for its technology, would protect American farmers when our budding genetic technologies flourish overseas, Robinson said. If farmers in foreign lands were to save the seeds from genetically modified plants, they would avoid the "technology fees" that biotech companies collect from North American farmers and gain a competitive advantage. "If we don't protect how that technology is going to be delivered to other countries, then we are, in effect, unbalancing the scales," he said.

Conversely, if companies don't have to worry about their genetically modified seeds being filched, they would be more likely to export them. So TPS was a good deal for all around, I was told.

If things worked out as expected, Delta and Pine Land would control the marketing of the Terminator, which could be licensed to biotech companies for a fortune. (Claiming its role has been to assure wide availability of the benefits of its farm inventions, the government has typically not taken its share of profits.) So Robinson had more at stake than the well-being of farmers. The early opposition popping up on the Internet and in the British press did not share his confidence. "There are those who want to frighten people and who say that the world as we know it has ended, and that we are going where nobody has ever been before," he said dismissively. Storm clouds were gathering, but it was hard for his company to see them down on the Mississippi Delta. I concluded that the company had not yet figured out how to present its new invention to the world.

As I left the old plantation, I snapped Robinson's photograph, and it would accompany my newspaper story. We talked about the budding sex scandal in Washington involving the president of the United States and a former White House intern. He tried to assure me, before bidding me farewell, that his company had the best intentions. "We're really pretty good people, and we're trying to do good things for people," he said.

Robinson told me something more important than I knew at the time: He said that his company's new technology might be attractive to Monsanto.

But in the flush of discovery, neither Robinson nor the United States government imagined that their best scientists were about to be foiled by a high-school dropout.

RAFI AND THE GENE GIANTS

In the early 1960s, a young Canadian, Patrick Mooney headed to Europe for adventure. It was the thing to do in that disaffected generation, especially if you were about to flame out of your Winnipeg high school.

Mooney's journey took him to Rome and, by a friend-of-a-friend connection, to a youth conference sponsored by the United Nations Food and Agriculture Organization. Somehow, he snared a paying job, as youth consultant on North America for the FAO director general.

"I had long hair and beads, and I had dropped out of high school. I was the perfect candidate for a U.N. job," Mooney recalls.

In 1977, his freshly cultivated sensibilities about the plight of farmers and his distrust of corporate power led to his cofounding an organization called Rural Advancement Fund. Seven years later, it became the Rural Advancement Foundation International, known widely as RAFI.

Over the years, RAFI blew the whistle on corporate consolidation in industry and the growing trend in agribusiness, particularly in the seed industry, of patenting genetic material. Speaking for small farmers of the world, the group earned a place at the table in United Nations gatherings. They were earnest folks not given to confrontation—Mooney never has been a fan of Greenpeace—and they presented their case in laboriously detailed reports they called communiqués.

Until the late 1990s, RAFI was, despite its industriousness, another do-gooder, non-governmental organization. Then came the Technology Protection System which, thanks to Mooney's momentary flash of inspiration, propelled RAFI into the limelight.

It was Hope Shand, another RAFI cofounder, who noticed an item in the *Wall Street Journal* about the TPS patent award. Operating out of her home-office in Pittsboro, North Carolina, near her alma maters of Duke and the University of North Carolina, Shand tracked down the discoverers and also phoned the Agriculture Department. When she asked a spokesman if the government had fielded any calls from critics, he didn't understand the question. Why would anyone protest?

A few days after the patent was issued, Shand sent out a news release that roused neither the press nor like-minded advocacy groups. Meanwhile, Mooney had just returned home to Canada from Mexico, and over the phone Shand brought him up to speed on the patent's scope.

Pat Mooney swears he never has seen any Arnold Schwarzenegger movie, let alone his portrayal in *Terminator* and its sequel of the T-1000 liquid-metal robot sent from the future. "But I knew what the Terminator was—something that comes in and stops something, terminates it with killer force," he recalled.

"We couldn't figure out why everybody wasn't reacting. I just said to Hope, 'reissue what you wrote, and call it Terminator.' So she slightly rewrote the same news release, and that made all the difference. Soon, everybody was talking about the Terminator."

In my recollection, *soon* was a matter of months. But by late spring and summer of 1998, when I was writing from Europe about biotechnology, I was hearing the word frequently across the Continent. Language made the difference. A metaphor as apt and rich as the Terminator might come along once in a decade. Mooney, brilliantly or luckily, had plucked a word that lay fresh in public consciousness thanks to the Schwarzenegger movies.

"I think that for a lot of urbanites, it just seemed grotesque or immoral," Mooney said long afterward, analyzing the course of the debate and the impact of his terminology. "Here was a technology whose sole purpose was trying to stop things that happen naturally—not just trying to stop something that had been happening for twelve thousand years."

For journalists, the word "Terminator" enabled the neat summation of a complex chunk of science almost too daunting to handle. There is no way that my first story about the new technology would have run on page one above the fold, as it did, had I used the words "technology protection system" rather than Terminator. For critics of the technology, it made an astounding difference. No human safety threats from modified food had been strongly implied, let alone proved. Environmental consequences such as outcrossing and potential resistance to Bt were hard to explain. Suddenly in this maze of probability theory and invisible processes, they had something to hold on to.

People could grasp the Terminator: It was a device by which corporate multinationals could change the way farmers around the world had operated for millennia, and it seemed to contradict what the companies were saying about the bounty biotechnology would spread across the planet. "I think that even people who don't follow agriculture instinctively got it," Shand said. "It shattered the myth that commercial biotechnology aims to feed people, which was their mantra."

The Terminator put flesh on the bones of a political argument—that a handful of life-science companies were bent on controlling the food chain. That argument was buttressed by a frenzied period of consolidation in the seed business.

Life-science companies were gobbling up the world's seed companies like starved dogs. The big players in the biotech industry understood that seed companies were the means of delivering their new genetic traits to the world's farmers. DuPont had purchased Pioneer Hi-Bred, the prize of them all, and Hybrinova S.A. of France on the way to becoming the world's largest seed company with $1.8 billion in annual revenues. Novartis—the product of a merger between Swiss giants Sandoz and Ciba-Geigy—had acquired or obtained an interest in more than a dozen seed companies active throughout the world, among them Rogers, a U.S. company; Hillshog, of Germany; Sluis & Groot, of the Netherlands; and Northrup King.

Dow's seed acquisitions included purchase or new holdings in two U.S. companies, Mycogen and Phytogen; three Brazilian companies, Dinamilho Carol Productos, Hibridos Colorado, and FT Biogenetics de Milho; Morgan Seeds of Argentina; and a 35 percent stake in Verneuil Semences of France. AgrEvo, which itself would be merged into the life-science giant Aventis, acquired Nunhems, a Dutch seed company, and Granja 4 Irmaos of Brazil. Zeneca, which would become AstraZeneca in a merger, had purchased Garst Seeds and VanderHave, among others.

Later, Syngenta was formed from the merger of Novartis and Zeneca, while Aventis was the name of the new entity merged from AgrEvo and Rhone Poulenc.

None of the "Gene Giants"—another RAFI-coined term—was more active than Monsanto, which spent billions of dollars in a buying spree that would establish the St. Louis biotech leader as the second-largest seed company in the world. Monsanto's acquisitions include Cargill's international seed operations; Plant Breeding International of Cambridge; Holden's Foundation Seeds; Corn States; Asgrow/Hartz; HybriTech; Stoneville Pedigreed Seed Company; and Sementens Agroceres of Brazil.

Then Monsanto made an announcement that would propel the Terminator debate to new heights. On May 11, 1998, the world's leading biotechnology company declared that not only was it buying DeKalb Genetics Corporation, the nation's second-leading seed corn company, but that it was acquiring Delta and Pine Land for $1.9 billion.

Even financial analysts were stunned. "Monsanto is creating a giant tollbooth in front of the cotton market and the soybean market and the canola market and the corn market," said Salomon Smith Barney's James H. Wilbur.

MISSISSIPPI GAMBLING

Along the Mississippi Delta, the Mississippi River deposited over the centuries earth so rich that entrepreneurs called it chocolate-covered gold. Mississippi was especially fertile, out-producing any other cotton fields in the world. The delta is rich in cotton and it is rich in blues; in music lore, the intersection of Routes 61 and 49 in Clarksdale, Mississippi, is the crossroads where guitar legend Robert Johnson sold his soul to the devil in return for success.

Cotton farmers in these parts who trusted their fields to the early magic of genetic engineering worried that they had made a Faustian bargain of their own. Randy Talley was wondering as much when I stopped to see him in the town of Bobo, where I visited in search of both farmers and one of the most fetching newspaper datelines I could recall. Talley was waiting for me near sundown, sipping whiskey on his porch, a shotgun at the ready.

Genetic engineering swept through cotton fields more swiftly than in any other crop, and if and when the Terminator becomes a licensed weapon, I expect it to first be deployed by cotton growers. As recently as 1995, no commercial cotton sprouted anywhere in the world. By 1998, 30 percent of the twelve million acres of cotton in the United States grew from genetically modified seeds, and by the new century, nearly two-thirds of the American cotton crop was engineered.

We don't eat cotton, so growers sowed the new technology largely unimpeded by health and safety concerns. But insects eat it; boll weevils, boll worms, tobacco bud worms, and, in some fields, spider mites, fruit feeders, and what farmers call tarnished-plant bugs and clouded-plant bugs. Several years into the era of gene-altered farming, no public benefit from the technology measured so bright, so undeniable, as the reduction from Bt cotton in insecticides sprayed in the southern United States, where cotton is king.

"It keeps me from having to handle a whole lot of high-powered worm poisons that we've used all our lives," Mike Williamson, a cotton farmer from Water Valley, Mississippi, told me.

Growing cotton in Mississippi has been a gamble ever since the Percy family began carving crop lands out of waterfront jungle in the middle of the nineteenth century. The costs of pesticides and fertilizer are staggeringly high, over $500 an acre, compared to $200 for corn. Cotton is especially vulnerable to the vicissitudes of weather.

So when Randy Talley heard about seeds engineered to, in effect, produce their own insecticide, he rolled the dice. He could only shake his head when he recounted to me what he saw in 1997 on much of the 1,350 acres where he planted the new anti-insect seeds that he insisted were faulty. "Green bolls were lying on the ground everywhere," he said of the pods that yield cotton. "I thought for sure I'd be plumb broke."

Talley didn't go broke; he bounced back thanks to a settlement and better cotton crops in the future. Walking his Bobo farm, he taught me a lesson about the nature of farming and the likelihood that the Terminator, like genetic manipulations before it, would never be stifled. Still recovering from the bad times he blamed on bad modified seeds, Talley was preparing for his annual gamble of planting a new crop of cotton. When I asked him if he would consider planting genetically modified seeds again, he paused only briefly before answering, "I might."

"DEMON SEED"

From the outset—even before the Justice Department approved the acquisition—Monsanto and the Terminator were a combustible mix. Monsanto already owned Stoneville Pedigreed Seed Company, which meant that if the Delta and Pine Land acquisition succeeded, Monsanto would control nearly 85 percent of the cotton-seed market in the United States. The prospect of such vast holdings fueled fears about agribusiness consolidation. Already, Washington was smelling the smoke.

In a "white paper," Monsanto wrote about concerns about misuse of the potent technology born in the Lubbock lab. The company said that it "welcomes the development and responsible use of seed-protection methods in general and is in full agreement with the principles that underlie the [gene-protection] technology."

Hope Shand was flying from a conference in Bratislava to Prague when she read the news of the pending merger in the *International Herald Tribune*. She fairly sprinted from the plane to get on the phone to anti-Terminator strategists. RAFI and a host of advocates were all over the merger at once. "Alarm bells started to sound," she recalled. "People had never heard of D&PL, but with Monsanto, they understood that this was a serious threat with moral implications."

Monsanto's high-sounding talk failed to convince the news media, among others. In this critical period in biotechnology's evolution, Terminator's potential for protecting intellectual property and reining in wayward genes got short shrift. Coverage was global in scope and savage in tenor. People in developing countries got their introduction to genetic engineering through the Terminator.

Under the headline "New Seed Technology Threatens Third World," an African news agency began a dispatch with these words: "A technique that disables a seed's ability to germinate when planted was widely criticized by experts as it could spell doom to agriculture in the developing world." African delegates at a United Nation's gathering on plant genetic resources requested aid in fighting biotech companies.

"A Demon Seed," read the headline on an Asia Intelligence Wire dispatch. In India's news media, developments in the Terminator were reported with alarm, feeding nationalist disdain for anything perceived to corrupt culture. An article in the *Hindu* began: "Ever since humans started farming ten thousand years ago, they have followed a basic tenet: save seed." That lead rang true with readers in India, where three-fourths of farmers save seeds for the new planting. The India Council of Agricultural Research, no enemy of biotechnology, insisted Terminator technology be banned.

In the United States, news media had all but ignored the expanding acreage of modified crops and the budding opposition. Now reporters found a "hook" for articles about the complex issue, and their stories in turn helped form the popular perception of inequality inherent in genetic engineering's magic. An Associated Press article out of Washington began: "A new technique that makes seeds sterile is sowing controversy among critics who say it will protect big-business profits while unfairly ending the age-old farm practice of saving a crop's seeds for next year." The *National Catholic Reporter* started a report similarly: "A new genetic technology that alters plant seeds so they are unable to reproduce will hurt poor farmers and has the potential to exacerbate world hunger, according to activists concerned with the impact of biotechnology and genetic research." *Time* magazine spread its report over two pages with "The Suicide Seeds" in big, bold type over this teaser: "Terminator genes could mean big biotech bucks—but big trouble, too, as a grass-roots protest breaks out on the Net."

That Internet campaign flourished. Thousands of people from more than sixty countries sent e-mails to the Agriculture Department demanding an end to the research. An early casualty was Monsanto's budding relationship with the Grameen Bank in Bangladesh. The Grameen Bank and its managing director, Muhammad Yunus, are known for their policy of microcredit—lending to small-acreage farmers who might otherwise not qualify for loans. In the spring of 1998, Yunus announced formation of the Grameen Monsanto Center for Environment-Friendly Technologies in Dhaka. With a relatively modest 150,000-dollar investment, Monsanto would be spreading conservation techniques that farmers in Bengali villages sorely needed— and potentially sowing genetically modified crops.

Pat Mooney was quoted widely as saying the arrangement "could turn poor but independent farmers into poorer and dependent peasants." The attacks on Monsanto and the respected Yunus in the summer of 1998 were fierce and unremitting, and the deal was canceled.

CORPORATE INERTIA

By 1998, biotechnology and its pathfinder, Monsanto, were losing ground in the struggle in Europe for consumer acceptance and European Commission approvals of products. Now, after announcing acquisition of Terminator, Monsanto was again flogged around the world.

As a political reporter for twenty-five years, I had enjoyed front-row instruction in the art of damage control. I'd seen it work for political leaders and would-be presidents, Bill Clinton and Dan Quayle among them. I'd watched it fail, most memorably riding the campaign plane in 1987 as White House hopeful Gary Hart's dalliance with a young woman became white-hot news.

As the pressure built, Monsanto attempted its own damage control. Inside the company, epic conference calls were logged to figure what to do. Hard-liners argued that backing away from the technology would constrain research. There was concern, too, about muddying the acquisition. In negotiations, Delta and Pine Land considered Terminator a commodity of great value. How would it look for Monsanto to declare that it wouldn't be useful to them, thereby diminishing the value of what they were acquiring?

Politically savvy operators in Monsanto wanted to take action. It was, in the words of one company strategist, a "burgeoning and uncontrollable controversy. There was a strong belief that you don't want to walk away from it, but there was also a recognition that this thing had the potential of doing great damage to the overall technology."

Meanwhile, extraordinary pressures were fragmenting the company. In June 1998, Monsanto had announced a merger with American Home Products, a deal that would create a twenty-three-billion-dollar company. But the companies couldn't agree on details of the arrangement, including the leadership structure, and the merger was canceled. Also pending was the Dekalb acquisition, which at $4.4 billion was bigger and more important than the Delta and Pine Land deal.

In Europe, the front line of the global war over genetic engineering, company officials were debating a fateful decision on advertising.

So at Monsanto, Terminator was just another ruckus, albeit one with extraordinary public-relations peril. Months passed, and the company did nothing, even as the voices of opponents grew louder and echoed in unexpected quarters.

The Agriculture Department became paralyzed, too. Negotiations to give Delta and Pine Land an exclusive license for the technology fell off the table, and the government decreed that its scientists needed special permission for Terminator work. A memo written to Agriculture Research

Service scientists in September 1998, described the Terminator as "a sensitive issue that requires an extra level of review."

In October, scientists and farm economists from around the world voted at the World Bank in Washington to condemn the Terminator. They were members of the middle-of-the-road Consultative Group on International Agricultural Research, which is committed to battling hunger. Seldom did this organization enter the domain of politics. Ismail Serageldin, a World Bank vice president who oversees the group, reminded me how the debate over Terminator had shifted to its threats to poor farmers. "We're talking about people who live on the knife's edge of subsistence. These people's very existence depends on their ability to grow the next crop," he said.

Finally, six months later, in April of 1999, Monsanto had something to say. The company declared that it would not market the Terminator until studies examined its environmental, economic, and social effects.

"Companies are researching these technologies because they believe they may provide a number of benefits, the primary benefit being protection of the investment required to develop the seeds. Such protection encourages more research and investment in future agricultural improvements and thereby would expedite access to the benefits of biotech seeds by farmers who want them," a statement read.

"At the same time, however, the fact that there is so much concern being expressed about this type of technology indicates that there are many who have serious misgivings about them and their potential impact on food production. We believe that the concerns about gene protection technologies should be heard and carefully considered before any decisions are made to commercialize them."

The Terminator would not soon be germinated, but neither was it terminated.

For Monsanto, it was a sobering episode at a critical moment in the company's drive to hold back the opposition. The Terminator was technology of the future, albeit a powerful technology. But it became a symbol for the present, a means to organize farmers around the world and a weapon for critics to bludgeon a paralyzed company.

E-MAIL ANALYSIS

My coverage of the global biotechnology debate won little applause from Monsanto, institutionally at least, or from its ardent defenders. One of the company's allies was Peter Raven, director of the St. Louis Botanical Garden and a renowned scientist who, in 2000, was awarded a prestigious National Medal of Science.

Raven was married then to a Monsanto official, Kate Fish, and the company generously supported the St. Louis Botanical Garden. When Monsanto's troubles began to boil, he became the object of criticism for his Monsanto connections. In April of 1999, Raven sent an e-mail to Cole Campbell, then-editor of the *St. Louis Post-Dispatch,* and to me, taking me to task for what he regarded as my criticism of Monsanto's half-a-loaf announcement on the Terminator. In an e-mail exchange that followed, I had the opportunity to play political consultant when Raven asked what I would have done.

Peter:

I hope that I would have been attuned to the warning signs more than a year ago when RAFI began to score on this issue. The European press was beginning to seize on Terminator; in European debates and conferences, Terminator swiftly rose on the list of concerns cited by critics. I saw it happening. Why? Because in a welter of abstractions and murky threats, Terminator is understandable; graspable. From my travels in the past year, I would say that it's hard in this broader debate to overstate people's fear of being controlled by multinationals. It's far more than "protectionism." Terminator is the perfect symbol for that argument. And the name is so deadly.

By early summer [1998], I would have said: "We gotta do something." Then, when I saw what transpired with the Grameen Bank and the power of the Internet for mobilization in the developing world, I would have said: "We gotta do something yesterday."

In my understanding, by the time Terminator could be commercialized, it will be passé and other methods to control gene expression will have evolved, probably in Monsanto's laboratory.

With that in mind, I would have said, perhaps even had Mr. Visionary Shapiro [Monsanto chief executive officer Robert Shapiro] declare in a public forum: "We have heard your concerns. We assure you that you would have nothing to fear from Technology Protection System; in fact, there could be some real benefits for small-acreage farmers. Nonetheless, we have decided as a company that it is so important to have your trust that we are announcing today that we will not use TPS. If and when we become the patent-holder, we will convene a panel of academic experts to help us decide whether or not there may be worthwhile public uses for this application. Along the way, we will be seeking your input. Why are we taking this step? Because we have heard you and we understand that food biotechnology needs your acceptance in order to yield its many benefits down the line."

Next day's headlines: "Monsanto Terminates Terminator."

In one deft swoop, by late last summer or early fall at the latest, critics and skeptics would have had Terminator removed from their arsenals. The line of demarcation between the United States and the rest of the world might not have become so pronounced. In short, the issue never would have festered.

ENTER DR. CONWAY

Inside Monsanto, too, sentiment persisted that the company must go further. As a Monsanto executive of over twenty years standing told me later, "If we started out trying to fuck something up, we couldn't have done a better job."

Gordon Conway felt the same. Unlike me, Conway was someone Monsanto paid attention to. Conway, who is British, is president of the biotechnology-friendly Rockefeller Foundation, which by 1999 had sunk $100 million into research on genetically engineered crops. An agricultural ecologist who worked for three decades in development programs in Asia and Africa, he wrote an authoritative book, *The Doubly Green Revolution*, arguing that the world's expanding population can be fed only by combining conservation with increased productivity.

In June 1999, during a meeting at a Washington hotel with Monsanto board members, Conway blistered the company. He told them that the industry was rushing products to the market without sufficiently evaluating their effects. He told them that rather than financing a new offensive authored by a public-relations firm, (as was about to be done), they should be developing with the public a "new relationship based on honesty, full disclosure, and a very uncertain shared future."

He advised Monsanto to disavow the Terminator, once and for all. When we chatted after the meeting, he recalled his experience in Thailand, where farmers planted thirty to forty different rice seeds that they often traded with one another. What if subsistence farmers ended up inadvertently with sterile seeds? he asked.

Monsanto officials wished their board had invited someone else to its gathering. Even more than at what Conway told them, they were irked at his news release announcing what he had said. The company issued its own statement using the Washington "frank and productive" argot to describe Conway's appearance. But the release said nothing of substance. "We will continue to reach out to people like Professor Conway and institutions like the Rockefeller Foundation to discuss the challenges and opportunities of biotechnology applications in agriculture," it read.

Five months later, Monsanto replied to Conway—and to the world. The company's chief executive officer, Robert B. Shapiro, said in a statement: "Though we do not yet own any sterile seed technology, we think it is important to respond . . . by making clear our commitment not to commercialize gene protection systems that render seed sterile."

Finally, after seventeen months of merciless clobbering for a technology it didn't yet own, Monsanto terminated the Terminator.

In Winnipeg, Mooney said he was happy at Monsanto's action. "This is a genuine move on their part," he declared.

Down in the Mississippi Delta, cotton executives seethed. Harry Collins faxed me a tersely worded statement. "Over the past year, negative comments regarding TPS have misrepresented the technology . . . Delta and Pine Land Company intends to proceed with its commercialization agreement with the [Agriculture Department] and we will continue to support TPS research."

IT LIVES

The biggest cotton company and the biotechnology leader never consummated their union. After Monsanto merged with Pharmacia & Upjohn in December 1999, the newly constituted company called off the acquisition of Delta and Pine Land. It wasn't in the plan.

A month later, Delta and Pine Land sued Monsanto for one billion dollars in damages plus one billion dollars in punitive damages. Monsanto, Delta alleged, was "dilatory, obstructive and unreasonable" in obtaining Justice Department clearance for the merger. Monsanto said it didn't appreciate being sued for two billion dollars by "a technology partner we've worked with for years."

Like the cyborg of film fame, the Terminator lived on in the laboratory. In the summer of 2000, I telephoned Harry Collins to see how things were faring. Terminator's ardent opponents liked to refer to him as "Dirty Harry," but he struck me as a decent fellow who bore up well under a challenging task—defending what the world saw as a freak. Terminator remained a ways from hitting the market, he said, but the genes for the system had been successfully engineered into tobacco.

Soon thereafter, Terminator's other creator, the Agriculture Department, had to fend off its new Advisory Committee on Agricultural Biotechnology. After six hours of debate over two days in committee meetings in July, eleven of the committee's thirty-eight members signed a declaration advising the government to abandon its patents.

"We are steadfast in our view that USDA's continued association with the Terminator patent is a fundamental mistake because the technology is potentially dangerous to biodiversity and global food security. Terminator technology has only one primary purpose—to allow private companies to exert greater control over the seed markets and extract more income from farmers forced to buy their products on an annual basis," their statement read.

But the majority of the committee did not sign—testimony to the resilience of a technology as hard to kill as its namesake.

On February 18, 2001, a fire in rural Visalia, California, at a Delta and Pine Land storehouse barely made the local papers. But an e-mail message sent out five days later by someone who claimed to be part of the Earth Liberation Front made the isolated blaze a topic of newspaper coverage from Portland to New York.

"We chose this warehouse because it contained massive quantities of transgenic cotton seed in storage," read the message, according to reporting by the *Sacramento Bee* newspaper. "But now, this seed will no longer exist to contaminate the environment, enrich a sick corporation, or contribute to its warped research programs."

It was a chilling message that went so far as to describe the method of the attack. "After cutting through a padlock on a door to get into the warehouse, we placed four five-gallon buckets filled with half gasoline and half diesel in strategic locations," the group claimed. "Windows were broken to provide the fire with oxygen, and timers were set. Within just a few minutes, the operation was complete."

Two months later, the industry received another ominous warning: The Ruckus Society announced that it would hold a week-long "Biojustice" activist training camp in southern California just before a Biotechnology Industry Organization convention in San Diego.

When I telephoned the Delta and Pine Land offices in Mississippi a few weeks later, the company held fast to their policy of refusing to comment about the fire or the shadowy organization that claimed responsibility for setting it. One official told me that he was loathe to say anything that might give credit to the saboteurs.

A message accompanying the claim of responsibility suggested that the biotech industry had more to worry about. "This action by the ELF comes after a quiet winter of no direct actions against genetic engineering. It is expected that with the upcoming growing season direct actions against facilities producing and testing genetically engineered organisms will resume."

PLANTINGS FOUR

I am no farmer, but neither am I a rookie when it comes to making things grow. In my kitchen hang ribbons from the Calvert County Fair won in the 1980s for odd peppers: Bolivian Rainbow, Poblano Rojos, and gnarly, nameless purple devils that I brought back from somewhere exotic. I have since forsworn smuggling seeds.

Looking back, I probably won not because of my garden skills but because the judges had never seen the likes of my weird entries. These days, you can find peppers like these in garden stores, just as you can find whole sections of murderous hot pepper concoctions in every supermarket.

My friends who are untroubled by genetic engineering make this point. They say that it is foolish to worry about altering ecosystems because so little that sprouts in civilization these days is native. Most of the nursery plants people buy originated far away, often in Central America. I look around my own neighborhood and see a few native azaleas and rhododendrons. I also see Japanese maples, Japanese black pines, lace bark pines from China, and snowball trees from the Middle East. I reside a hundred-yard dash away from an overgrown dip we call Kudzu Valley.

In my own yard, I have an exotic Costa Rican elephant ear (that came from a fancy garden store) and Yemeni garlic (from my backpack after a trip to the Middle East, before I took my no-smuggling vow).

Unless we're one of the five places on earth where the growing of food originated (Southwest Asia, China, Mesoamerica, Andes-Amazonia, and the eastern United States), we're eating food that came from someplace else. I happen to live in the eastern United States, but I'm not growing any of the few plants believed to have originated here, among them the goosefoot plants, spinach and beets, that Native Americans grew here about forty-five hundred years ago.

I guess my friends have a point.

Amateur that I am, I am surprised at how swiftly my soybeans have leaped from the soil. But I know that all the rain we've had gets the credit, not my skill. Already, my genetically engineered beans and the conventional beans are eight inches tall. In the race, they are even.

But my soybean patch is weedy, so it's time to spray Roundup. The only weed killers I've deployed are my hands, so spraying herbicide will be a psychological challenge. It will be a challenge for the killing juice, too, because the rain has fueled the growth of all sorts of curious vegetation plus Virginia creeper, Japanese honeysuckle, and native poison ivy that attacks from my neighbor's unruly yard through my split-rail fence.

We'll see what this Roundup is made of.

8

AN ORGANIC CORNUCOPIA
IN THE GARDEN OF GLICKMAN

From what are we to protect Eden if not from our own work? The
more we work the earth . . . the more we are obliged to protect it.
—Evan Eisenberg, *The Ecology of Eden*

The discovery of a new dish does more for the happiness of mankind
than the discovery of a star.
—Jean Anthelme Brillat-Savarin
Nineteenth-century chef

DUE NORTH OF the White House, up Sixteenth Street and two blocks to
the east, the crack users, the brown-bag drinkers and the District of Col-
umbians who scuffle in the shadows of power awakened to new scenery
in the waning days of 2000: a glittering Fresh Fields supermarket with its
organically grown largesse. On its fifth day of operation, the supermarket
offered a stage for the power-wielders of Washington to announce the
new National Organic Standards for the food industry. The Department
of Agriculture was formally releasing a 544-page rulebook for organic
certification and, after ten years of bureaucratic shuffling, bestowing its
imprimatur—a boldface USDA label—on the wares of a burgeoning industry.

By mid-2002, organic growers and the retailers who sell their wares
will live under a code defining the class of foodstuffs produced without
pesticides, hormones, and most of the rest of the additives out of the

The bins of American farmers have filled increasingly with genetically modified grains since 1996, the year they were first planted commercially. But many farmers began to worry about their exports after the emergence of a global resistance to modified food.

chemical closet of modern, intensive farming—including genetic modification. Vegetables, fruit, meat, eggs, and dairy goods altogether untainted by chemical processes can rightfully refer to themselves as "100 percent organic." Foods labeled simply "organic" must have at least 95 percent organic ingredients. A third claim, "made with organic ingredients," can be made with products containing at least 70 percent organic ingredients.

In one of his last public duties before yielding his job to the new Republican administration, Agriculture Secretary Dan Glickman planned to arrive at Fresh Fields to announce the government's organic rules to the world—and to unveil the new brown-encircled green label proclaiming USDA Organic. To get to the show, Glickman had to wind his way from his office in the Whitten Building, on the grassy mall that stretches westward from the Capitol, through the fringe of Washington's bad-ass neighborhoods.

For families living in zones of decay, putting food on the table poses challenges beyond paying for it. Supermarket chains have closed stores in neighborhoods like these to follow population shifts to the suburbs. As a result, inner-city dwellers often make do with the unhealthful, processed foods and the inflated prices of convenience stores. In Washington, D.C., the menu of the Logan Square neighborhood expanded exponentially on the morning ten days before Christmas when Fresh Fields opened its brick-and-glass whole foods market.

Glickman's black Ford sedan turned onto P Street, passing the Mid-City Fish Market a few doors down from Fresh Fields. Rather than $1.89 Big Breakfasts and fish "sandwishes" advertised by the market's red-over-white hand-lettered sign, Fresh Fields promised its shoppers "Sparkling Seafood." At the new market, so the signs read, "we hand-pick our sea-food daily, right from the fishing docks . . . pure and natural." Fish here—including the fifteen-dollar-a-pound tuna steaks and the salmon flown in from Iceland—"is free of preservatives, sulfates and dyes."

At Best-In Liquors, a narrow, dirty-window store a few feet from where Glickman was disgorged from his car, patrons shove money through a revolving window to a silent woman behind bulletproof plastic for their sweet wine or a "forty" of malt liquor. But next door Fresh Fields, with aisles open wide and bright, offers locals Bacchanalic delights from shelves replete with cabernets and sauvignon blancs that can be touched, if not afforded. Beyond the promise of eating freshly and lavishly, people are invited, not so subtly, to mend themselves; to join the ranks of people who belong to the category advertised as Whole People, those who inhabit a Whole Planet.

For folks of healthy income or none at all, passing through these doors is entering a new Garden of Eden. Visiting the new whole foods markets is much like a trip into an open-air food bazaar, perhaps in Costa Rica or Italy, where nature's quirky bounty is piled high, ripe and rich, before farmers who've carted it in from their fields. If the growers aren't present at the Fresh Fields, their photographs are: Gary Whidden of Florida proffers his oranges and Dan Peixoto of California embraces a cabbage as big as his ample belly. Their photos, along with their vows forswearing chemicals and hormones, are reassuring; shoppers traversing the aisles can be persuaded by the testimonials and the pledges of purity implying protection from harm. They have escaped the territory of manipulated, altered, fabricated, abnormal food. In eating organic foods, they are becoming whole.

Shoppers run a gauntlet of placards in search of a potato or two. From preschool on, we're taught that food is nutritional; understanding how it is now "transitional" takes new thinking. So shoppers puzzle over the sign just inside the front door, above shining apples and pears the size of soft-balls. What's transitional, the fruit or them? Who knows? It might be a stop on the way to becoming one of the Whole People, part of the motto of the Texas-based Whole Foods supermarket chain, which advertises itself as the world's biggest organic retailer, with more than 120 stores. For enlightenment, shoppers read the small print beneath the new word. Tran-sitional, they learn, means "grown without the use of synthetic pesticides, fungicides or fertilizers. Land must be farmed with organic methods for a number of years before crops can be certified as organic."

So this transition business is not about people or food after all. It's about land and the time land takes to recover after having been soaked in chem-icals and fortified with various factory-made fluids. Not unlike some of the shoppers, I imagine, embarked on their 12-step programs. I wonder if the Whole People in the back room of this Washington, D.C., store made this sign with the clientele hereabouts in mind until, a few days later, I read the same words at a Whole Foods market in an upscale San Fran-cisco neighborhood—and see the same photos of transitional farmers.

Organic foods and the perceptions that surround them are huge business coast to coast, which is why Dan Glickman would stand in front of four-teen television cameras to declare rules for production. Under these rules, organic farmers selling over $5,000 worth of goods yearly and desiring the government's label would need accreditation. The rules require growers to rotate their crops to prevent erosion and to feed livestock only organ-ically grown grain if their meat, milk, or eggs are to be certified organic. There would be no government testing, though; an audit trail is the spine of the new system.

In the United States in the late 1990s, organic foods grew by more than 20 percent annually into a six-billion-plus-dollar business. The time had long since passed when organic food was the domain of tie-died sprout-crunchers. In Europe, which has experienced similar growth, organics were closing in on $5 billion in annual sales. Mainstream food manufacturers and their allies in the biotech industry are fond of saying that the organic trade is puny compared to the overall food industry, which in the United States alone exceeds $400 billion. That may be true, but the organic-food industry threatens factory-food manufacturing beyond the economics of the present.

In the past, organic agriculture was a pest that mainstream agriculture could afford to swat away. But as the new century arrived, fast-growing organics were poised to threaten Big Farming, particularly biotech farming, by hastening a two-track food system from farm gate to dinner plate. The prospect of this enormously costly separation—in grain bins, rail cars, and the ships of international commerce—was slowing the expansion of genetically modified agriculture. In addition, the inherent needs of organic farmers to have their land kept free of windblown pollen from modified crops was posing regulatory and liability questions that the United States government had not summoned courage to face.

On hand for Glickman's announcement at Fresh Fields was Katherine DiMatteo, who has watched her Organic Trade Association grow sixfold to twelve hundred members in a decade. "It's very difficult to coexist with genetic farming because of pollination," she told me. "And if there turns out to be impact on soil and microorganisms, there may be additional incompatibility." But it will take time, she said, to see just how much genetic "contamination" people will tolerate in their food. The Organic Trade Association, of which she is executive director, will demand a seat at every table where the rules for modified food and seeds are drawn, she said.

No wonder the biotech industry has fought back by warning of health threats of organic foods and by accusing the devotees of organic agriculture of practicing elitism in advocating farming systems that will not feed a hungry world. One company official told me that he forbids his family from entering whole foods markets, scorning patrons as "all the beautiful people who shop in their sunglasses."

On this day in Washington, D.C., the biotech industry had more tangible reason for worry. After the industry worked for most of a decade to persuade people that GMOs are the path to better health and a cleaner environment, the U.S. government was about to announce that genetically modified foods would be prohibited from proudly bearing the certified organic label.

"FULL-THROATED" ORGANIC DEBATE

Dan Glickman stood at a podium in front of organic foodstuffs arrayed in a gleaming metal bin of ice. There was bottled milk, yogurt, orange juice, tea, no-fat salad dressing, sesame tahini, and multigrain bread that requires vigorous chewing. Behind him stretched a panoply of foods unfamiliar to many people: jicama, taro root, and gnarly hunks of horseradish that resemble unearthed bones. There were all manner of chilies presented by their likeness of color and arrayed in shiny wooden baskets: cubanelle, Anaheim, serrano, and habanero. The Frank Zappa song, "Vegetables on Parade," began playing in my mind.

A decade had passed since Senator Patrick Leahy, a Democrat from Vermont, engineered passage of the Organic Foods Production Act. Back then the fledgling industry, which had yet to crack $1 billion in sales in the United States, sought regulations to prevent false claims about organic food and to deliver clarity as to what organic entails. The term was being thrown about too loosely for growers bent on forging a true and vibrant organic trade. But there was powerful opposition from the food and farm industries from the beginning; the legislation remained bound up in the House and had to be appended to the 1990 Farm Bill in order to escape the Congress. It was signed into law but languished in the Agriculture Department, which was loaded with people suspicious of and downright hostile toward farming that spurned chemicals and the other intensive agricultural methods government researchers had helped to invent.

Not until Glickman took over the department in the mid-1990s did the government get serious about carrying out its orders from Congress. That progress Leahy, now standing in the new supermarket's organically grown cornucopia, observed.

"We've had a ten-year gestation, and now we have the delivery," Leahy said, crediting Glickman. "There is a desire to have a choice, and the people who want that choice know that it is a real choice."

The effusive Glickman got caught up in the moment. "I'm a frequent shopper at this store," he said. I saw an alarm go off behind his eyes, a reflex conditioned during his years as a Kansas congressman. *You can't say that, Dan; they'll take it as an endorsement of organic foods—if not this store.*

"I buy both organic and conventional," he quickly added. (In truth, Glickman had never set foot in this newly opened market; he was recalling his experiences shopping at another whole foods markets in the region.)

By 1997, Glickman decided that the time had arrived to end the delays. But not right away. In a prelude to the formality of government rule-making—and recognizing opposition within the food, farm, and biotech industries—he requested public guidance.

What would you think, he asked, if food labeled organic were fertilized with municipal sludge? If it were treated with ionized radiation? If it were grown with genetically modified seeds?

Up to then, never had the Agriculture Department received more than eight thousand public comments on any issue. What occurred after Glickman's request for advice signaled the depth of sentiment on organic food and set a benchmark in American citizens petitioning their government. By the spring of 1998, the Agriculture Department had received 275,603 public comments on the proposed organic rules, most of them condemning the prospect of genetically modified food wearing the organic label.

Another 40,774 e-mails and letters poured in later in a firestorm of public comment that Glickman referred to at the Fresh Fields as a full-throated debate. "It was so full-throated," he added, "that I think it was choking me on occasion."

Two years before, Glickman had emerged as a voice of moderation in a government that promoted genetically modified foods in word and deed. "We can't shove it down the throats of the world," he had said, adding his voice to a widening debate. Still, he chose his words carefully, as he did when speaking of the good of organics.

"I think it's generally recognized that organic agriculture probably does provide some environmental benefits in terms of reducing use of certain things like pesticides. By saying that, I'm not implying that the food is any safer, more safe, less safe, or anything else. But I think that one of the reasons consumers have an interest in organic foods is because of the environmental benefits," he said.

Is that why you buy them, Mr. Secretary? a reporter asked.

The alarm was sounding again. "Frankly, they haven't put food in front of me that I haven't enjoyed," he said, laughing off the question until he realized he had to answer it.

"I know when I buy organic foods that they will have certain qualities that are different from nonorganic foods. But I tend to buy all foods. I'll look physically at what it looks like. But I think that organic offers an edge to some consumers and, in my case, it's offered an edge on occasion. But I don't buy it exclusively," he said.

Most of the questions to Glickman had to do with GMOs, formally banned in the new rules from carrying the organic label. The no–genetic modification rule was troubling to food industry representatives such as Tim Willard, spokesman for the National Food Processors Association, who showed up to watch the proceedings. Willard told me he believed that there would be an effort to reopen the rules to allow genetically modified foods to be classified as organic. "It's clear that in the future that there will be advantages in biotechnology that go to the heart of organic—pesticide reduction, no-till farming and other environmental benefits," he said.

But not yet, not under the rules that escaped the government bureaucracy under Glickman in the final weeks before the arrival of the George W. Bush administration and the potential of another cold shoulder toward organic farming. In one of his responses during questions from reporters, Glickman displayed his willingness to depart from what has been the script, in Democratic and Republican administrations alike, of insisting that its regulations governing genetically modified foods were unassailable. A Japanese television reporter asked Glickman about U.S. exports of the genetically engineered StarLink corn tainted with a protein that could produce an allergenic reaction, an explosive revelation that would echo long into the future.

After reassuring the reporter and his country that the United States would oversee proper testing, Glickman nonetheless took his own government to task, saying: "I think that whole issue shows us the way that some of these products are approved need to be upgraded."

THE GREENING OF GLICKMAN

Agriculture secretaries in America tend to be remembered when other cabinet members are forgotten. In the New Deal, Henry Wallace, a Confucius-spouting Iowa corn breeder, made his mark as a champion of food reserves and soil conservation. He also started the world's most successful seed company, Pioneer Hi-Bred, and in 1941 became Franklin Roosevelt's vice president.

Earl Butz's 1960s' exhortation to farmers to plant "fence row to fence row" helped to promote an era of farm prosperity while chewing up protective buffer land. Unfortunately for Butz, he is better remembered for telling a racist joke.

In the 1970s, Bob Bergland spoke more passionately than any government leader before him about the loss of family farms to agribusiness consolidation. During the deregulatory heyday of the Republican 1980s, John Block stood as an unlikely champion of conservation. In the 1990s, Mike Espy, a former congressman from Mississippi, became the first African-American agriculture secretary, a measure of justice after a history of government discrimination against black farmers. Espy resigned in scandal but was acquitted in 1998 of illegally accepting gifts and travel.

Glickman, the last agriculture secretary of the twentieth century, hoped to be remembered for ending government discrimination against black farmers. But he may be recalled for something else: triggering a reevaluation of a hard-edged U.S. policy in the world war over genetically modified food.

Glickman had fought as a warrior for biotechnology, trumpeting its potential to feed the world. He had championed the American government's position so faithfully in squabbles with Europe that detractors accused him of yielding to trade officials and forgetting farmers in his charge. Then he felt the sting of the gathering resistance to GMOs: In Rome, in

1996, he was pelted with seeds by naked protesters. But he held to the Clinton administration's no-compromise line on GMOs.

Finally, he broke ranks. On April 29, 1999, he made a speech at Purdue University that, though unnoticed by the national media, sent shudders through the industry. The United States, Glickman said, can no longer shove GMOs down the throats of the world. Since the mid-1980s, the biotech industry had received support from the mighty United States government from top to bottom. Suddenly, that support looked to be shaky.

A few months later, he ruffled more feathers when he asserted in a speech at the National Press Club in Washington that companies ought to consider labeling modified food. Those statements also irked government officials who lacked Glickman's sense of the global opposition. An aide recalled later that it was a rough period for Glickman in an administration that insisted the world view GMOs as it did.

Gradually, the American government's tone moderated. And, in my estimation, Glickman may be remembered more than anything else for forcing the government to change its tune and consider better regulations of genetically modified food. "This issue will be around for years and years," Glickman said in Washington in the spring of 2000 as he welcomed experts to an advisory panel he had called on biotechnology. "We as a society must sort through some very complex issues to make informed decisions that are in the best interests of all involved—consumers, farmers, processors, everyone in the food chain."

My newspaper stories were the first to report his shift. In interviews over two years, Glickman told me about the evolution of his thinking.

When I asked him in 1999 if he was trying to send a message to the administration by breaking ranks, he told me he was. "I was saying this shouldn't be a steamroller. You can't force-feed GMOs down people's throats. There's a growing concern about what people eat, what goes in their mouths. We have to address those concerns and we have to allay fears. We have to build confidence."

In his global travels, Glickman said, he was hearing many questions that had barely begun to sound in the United States. "Just because a product is out there, a certain technology, doesn't mean that people willingly accept it. There are certainly more and more questions being asked about biotechnology, and those questions must be answered. They cannot be brushed off. They must be dealt with. Otherwise, what will happen is that the consuming public, both here and abroad, will begin to believe that there are problems with it. Or that people are afraid to answer the questions. Or that, perhaps, there's arrogance that won't let the questions be answered. All through human history, progress has been made. You can't stop progress. But you have to recognize that concerns have been raised and those concerns have to be dealt with," he said.

Glickman said that he had begun to reevaluate where the United States stood not long after being pelted with grain in Rome. "I came back and I asked, 'what the hell is going on?' Because at this point in time, the word was that this was the technology of the future. This was the moral thing to do. This was going to be the only way we're going to be able to feed the world. I thought, 'what is wrong with these people?'

"I still believe that the technology will produce an agriculture that will be able to feed the world in a more sustainable way. But it doesn't mean that it's written on Mount Sinai or that there aren't questions that have to be answered. That's the era we are entering now," he said.

There is no way to stop the public policy debate emerging in the United States, Glickman said. "There are many kinds of science. There's physical science. There's political science. There's also social science. There are all sorts of ways you have to deal with people, and sometimes, the physical scientists only see their area. And sometimes, the social scientists only see their area. We have to blend all of these. Because, ultimately, if the consumer doesn't buy, the technology isn't worth a damn."

A few days before Glickman left office, I found him laughing. When I turned up at his office, he was practicing for his last major appearance as a cabinet member, an engagement unusual for the nation's overseer of farm policy. Glickman would be speaking at a National Press Club lunch about humor in public life. As part of his repertoire, he was preparing to make light of his experiences with GMOs, relating his story about being pelted with grain by anti-GMO protesters. The portly, balding Glickman joked during his speech that women are moved to remove their clothes in his presence. "Danny," he recalled his late mother asking him after the episode, "what were those naked people doing to you?"

Glickman had a reputation as the jokester of the Clinton administration, the kind who would pass humorous notes in cabinet meetings like a sixth-grader in science class. But Glickman's gradual softening on biotech policy was anything but a laughing matter in the Bill Clinton White House, where the unwavering biotech promoters held sway. During our chat, Glickman admitted publicly for the first time that he had always worried that he might be going too far. In 1999, before a previous Press Club appearance, he hadn't, he said, submitted his speech on genetically modified food to the White House for the usual vetting by policy hawks and lawyers. He had worried that what would be left after sending it through the White House grinder would be sterile. What he said that day was anything but sterile, for he declared that the food industry ought to consider labeling their modified foods. Afterward, he was treated as a heretic.

"There were some people in this government who were very upset with me. Very upset. They thought that I had changed our trade policy unilaterally. I heard rumors through the grapevine that they were threatening [public] remarks about it. Some people agreed with me, but others thought that I hadn't properly vetted the speech and that I had cut the legs out from under our trade people. Like any politician, I wanted to be loved, so I was very nervous about it," Glickman said.

He was nervous until he bumped into Hillary Clinton at a state dinner.

"I liked the story about your speech that I read in *The New York Times,*" she told him.

"There were people in the White House who didn't like it," Glickman responded.

"I liked it," she repeated.

"I knew," Glickman told me, landing the punch line of his story, "that I wasn't going to be fired."

President Bill Clinton also eventually supported him, as Glickman said he judged by his assignment as *de facto* administration spokesman on biotech in Clinton's final months. "He's a good politician. He knew where this debate was going," Glickman observed.

In our conversation, Glickman spoke of the tension he had felt at being pressured by the administration and the industry to be a cheerleader for a technology that he was regulating. "What I saw generically on the pro-biotech side was the attitude that the technology was good and that it was almost immoral to say that it wasn't good because it was going to solve the problems of the human race and feed the hungry and clothe the naked. There was a lot of money that had been invested in this, and if you're against it, you're Luddites, you're stupid," he said.

"That, frankly, was the side our government was on. Without thinking, we had basically taken this issue as a trade issue and they, whoever 'they' were, wanted to keep our product out of their market. And they were foolish, or stupid, and didn't have an effective regulatory system. There was rhetoric like that even here in this department. You felt like you were almost an alien, disloyal, by trying to present an open-minded view on some of the issues being raised. So I pretty much spouted the rhetoric that everybody else around here spouted; it was written into my speeches," he said.

The other side in the debate also confounded Glickman. "There were some people at the extreme end of the environmental movement who were irrationally opposing this technology in the way new technologies have been opposed throughout history. They were the forces against change, whatever change was," he said.

Glickman believed, he told me, that the time had come for the United States to review the regulations that had been largely in place since the deregulation days of the mid-1980s. There needed to be more clarity, he said, adding a warning to his successor, Ann Veneman, that she could not expect to duck the global controversy. "Biotechnology is going to be thrust on her, as Dick Cheney would say, big time. Whether she wants it or not, it will be on her, like it was on me, big time," the comedian said, mimicking the voice of the new vice president.

9

ICE CREAM WITH EARL

The exploiter asks of a piece of land only how much and how quickly
it can be made to produce. The nurturer asks a question that is much
more complex and difficult: What is its carrying capacity?
—Wendell Berry, *Settling of America*

As I TRAVELED the frozen terrain of Iowa in the 2000 election season, pieces of the puzzle started fitting together. Hundreds of Iowa farms had disappeared since the 1980s, indirect victims of agribusiness mergers, consolidation, or expansion by land-hungry neighbors. Now a few of the Iowa farmers I met wondered if they had something new to fear: genetically engineered crops.

The tolling bell for the family farm has a familiar ring in these parts of America, and it rang long and loud during the farm crisis of the mid-1980s, when operations of all sizes collapsed under burdens of debt. But what was happening in Iowa in the winter of 2000 looked different. It reached beyond the pain of cyclical low prices for grain and livestock. It ran deeper than the shortcomings in the subsidy-cutting Freedom to Farm Act of 1996, which clipped holes in farmers' safety net.

"Did you know," a farmer in northern Iowa asked me, "that there are more people behind bars in this country than farmers working the land?"

The disappearance of family farms severs a bond between people and the land that has been revered since this nation began. Thomas Jefferson proclaimed: "Those who labor in the earth are the chosen people of God." When Harry S Truman left his family farm in Grandview, Missouri, for

Washington, his mother praised his farming abilities, not his political prowess. Her soon-to-be-president son could "plow the straightest row of corn in the county," she boasted.

I am aware of the tendency to romanticize farming. Is farming not like any other business undergoing mergers and shakeouts? Are consumers better off in the end? Are farmers as pure of heart as imagined by the piners for the good old days?

Talking to Earl Sime, of Radcliffe, Iowa, in the basement of the Iowa State Capitol one afternoon, enlightened me—and confused me. It was the opening day of the Iowa General Assembly, and Earl had joined farmers from across his state in showing up to say their piece. Earl had plenty on his mind.

At 265 pounds with a huge upper torso and hams for hands, Earl looked as though he could have played football at Iowa State University, where he went to college. Earl talked about farming like no one I'd met, and he was candid in ways that many in his business would find disquieting. His words poured forth between spoonsful of soft ice cream dished out by members of the American Legion. Earl talked about genetic engineering, about corporations, and about how farming works today. He had more answers than I had questions, and after a while, I just shut up, ate my ice cream, and listened.

"I know what a self-righteous, self-indulgent, sonofabitchin' Farm Bureau Republican is. The reason I know what one is like, the way one acts, is because I am one. I've belonged to the Farm Bureau for forty years, and I've been a Republican for fifty years.

"Why does a farmer use GMOs? He can use Roundup Ready and keep his fields clean as a whistle. If he's gonna lose it to weeds, he ain't gonna get much yield. Maybe he can improve his yield. He can come out of this like a fat roaster.

"Do you honestly think that this Iowa farmer—who's being stretched by his banker, stretched by his wife, stretched by his family and their needs, stretched by the fact that he wants to buy another new combine, stretched by the fact that he wants to have a clean field, wants it to look good to his neighbors and his landlord—do you think he

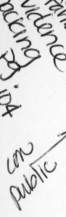

gives a frank damn about the possibility of this seed maybe damaging the health of someone in Europe who's gonna eat it? Or a chicken that eats it on the way to Europe? Do you think that farmer cares? That's a laugher. That's a laugher.

"Now he does care about his markets. I've never planted Roundup Ready myself. I don't want to spend twenty-one bucks for technology. I'd rather be a pirate. They talk about pirating seed. If you raised 'em last year, put 'em in a bin and use them this year. Then they come and hang you for it. More power to the pirates is what I say. I used to care and give a damn and put the most favorable construction on things. I've become a little callous over a period of time.

"I believe I'll have some more ice cream."

"Big corporations don't give a frank damn about the family farmer. Corporations will have you from stem to stern, from ship to shore. There will be control, control, control. There will be control from your banker's standpoint. You've got a marriage between seeds and chemicals. The whole thing is tied together. Do you think you'll own it? You'll just pour the sack of seeds, and you probably won't even own the sack. The only thing you're going to be is a tractor jockey. Just a tractor driver. They'll own the final product. They know the technology is in the seed, and they're gonna own it. If it's good stuff, and it's worth bucks, they'll own it.

"If there is a technology in, say, a certain kind of corn to be fed to certain kinds of pigs that are sick—they can put the nutritional kickers in the corn to straighten out, say, a pig's digestion. They'd be feeding it to a certain kind of pig in these corporate hog factories. Who do you thinks gonna own it? The sloppy farmer? Hell no.

"Do farmers understand what's happening to them? Many of them don't. But I'll tell you, many of them do, too. But it's like a storm; there's not a goddamn thing you can do about it if you happen to be in the middle of it.

"We're at quite a crossroads now, and I don't know how many people see it. Not as many as should, for sure. Corporate took over the chickens twenty-five to thirty years ago. Corporate has taken over the beef. Corporate's taken over the hogs. Next is the land. That's just ready to flop. Everything is falling into place big time. DuPont and Pioneer are going together. Monsanto and everybody else are going together. All the traditional name plates we've been proud of are either out of existence or they're one of four, one of six, one of twelve.

"They're looking for a farmer that can farm and lose money and have three other jobs to put beside it and have a wife with two other jobs. And you put that all together and say, well, I made some money. You're not talking farming here. You're talking business in several arenas. Maybe your wife is head nurse here at Mercy Hospital in the heart division. She's probably getting paid more in one year than you've ever seen in a decade on the farm, if you've had a little tough luck. Maybe you're a seed corn salesman or a school bus driver. It's as plain as 'do you have tires on your tractor?' No, I drive my tractor without tires. I just drive on the rims because I found it's cheaper. Because that way, I don't have to buy tires.

"You want some more ice cream? I believe I'll get me some more."

–––––––––––

"The farmer today doesn't give a frank damn about his neighbor over there farming. The only thing he cares about is if he can own his farm. If my neighbor kicks the bucket, then by God, I'd buy his farm. I'd buy his machinery. That's the story of agriculture today, so help me God.

"My dad was born in 1894. My dad farmed 240 acres with his three brothers. But he didn't spend one-fifth of one-tenth of one second thinking that he should own part of his neighbor's farm or own more land.

"It's all economies of scale now. If I died, somebody would say, 'Earl farmed it with all that fuckin' junk. Now I can farm it with what I got, maybe just add one more piece of machinery.' There's jealousy and greed and conniving from stem to stern. And I'm sorry to say, that's what it's all about. That's it, period, end of sentence. That's why you've got farmers saying it ain't fun anymore.

"But it's a beautiful life. We've got cabs on our tractors. We've got heaters in our tractors. We've got windshield wipers with water to spray. We've got flashers for goin' down the road. We've got fuckin' turn signals for turning in whatever direction we want. We've got power brakes. We've got radios and stereos. You can stick in a CD or a tape. We've got television in the cab. So you got anything you want nowadays.

"It's a sad situation. Agriculture is truly in a sad situation.

"Once it's over, we can't go back. Why? No grandfather-father-son thing. No uncle-nephew thing. If you're gonna be a farmer, you're gonna have it in you by junior high. We're talking about somebody growing up in farming. They're riding along on the tractor. But you lose a generation, and it's gone. And it won't come back.

"Man, that was good ice cream."

I telephoned Earl a few times over the next year to find that his views on the precarious state of the family farmer hadn't changed. The trend is bigger, bigger, bigger, he told me in the spring of 2001.

"Agriculture is in a sorry mess. In 1975, my [grain] elevator served one town and my other elevator two towns. Now one of the elevators is nine towns and the other is ten towns. Does that tell you anything? Does that suggest we'll be having small farmers in the future? The small farmer doesn't have a snowball's chance in hell.

"Yeah, farmers are thinking about GMOs and that Bt stuff. And thinking about Mad Cow Disease. But you know what a farmer is really thinking about right now? How in the hell am I gonna get my nitrogen [fertilizer] paid for?

"Is the new farm program gonna bring back family-farm agriculture? Why, hell no. And we in this nation don't give a tinker's damn. We don't care if chickens are in a factory barn with a hundred twenty-five thousand chickens. We just want eggs. It's better than a case of eggs being sold out of the back seat of a car. We don't give a damn. Otherwise, there would be fighting in the streets.

"You can write a nice story about the family farm, but it's gone."

PLANTINGS FIVE

At a neighborhood party, I let it slip that I am growing genetically engineered plants. It must have been the wine, because I had not planned on divulging my secret. I did not anticipate the reaction.

Nor had I expected the unfamiliarity with genetically modified food—even from neighbors who read highbrow journals and hold biotech stocks. They are among the majority of Americans who, despite the ruckus in Europe, don't know a great deal about genetic changes in food or the players orchestrating them.

They're familiar with the organic/non-organic dichotomy; indeed, some of my neighbors stock their larders from the shelves of the whole foods market in Annapolis. By and large, they're pesticide-averse. My community is among those along the Chesapeake Bay that has formally forbidden entrance to the county trucks spraying malathion and related concoctions to snuff out mosquitoes.

But there's not much understanding of this new business of modified food or whether it's a good or bad thing. I find myself peppered with questions, and I think I must sound like a Monsanto pitchman. "No, there aren't any studies suggesting that it hurts you," I tell one young mother. "There have been reductions in spraying of insecticides, especially in cotton fields in the South," I say to another mother who had asked, "Why do people plant the stuff?"

Another woman asks: "What happens when we introduce all these new genes into the ecosystem? Do we know the long-term impact?" I find this conversation becoming too serious for a spring evening. The women, especially, want to engage me on this issue, and I try to change the subject to the availability of the Atlantic blue crabs, which some of my neighbors harvest for a living.

"Why don't we get to know on the label if the genes have been tinkered with?" I hear.

"Why did you say you planted this stuff?"

I use the old "need to fill my glass" escape. But at the hors d'oeuvres table, I hear my neighbors still talking.

"They're not any different from other plants," says a young fellow who does something with computers.

"Yea, but do we want them growing around here?" one of the women asks. She looks over at me and then looks away.

10

THE PIGS OF CHEZ PANISSE

Eating is an agricultural act.
—Wendell Berry

You can take a pig and put lipstick on it and call it Monique,
but it's still a pig.
—Former Texas Governor Ann Richards

A SHORT WHILE ago, Paul Gebhart returned from Peoria with a check for $946.06. At daybreak, he'd departed his farm, in Christian County, Illinois, for the 228 mile round-trip in a 1987 Ford minivan, the type of vehicle favored by parents taking a child to soccer practice. Maybe two or three kids. This day, as on most Wednesdays, Paul made the trip in the minivan accompanied by five three-hundred-pound hogs. They lounged on straw where a backseat used to bolt.

"They just ride along with me. I haven't figured out yet what kind of music they like," he says.

The pigs, Duroc and Berkshire crosses, are raised organically. When the minivan gets to Peoria, Paul slides open the door and the pigs climb out, like kids on to a soccer field. Soon they are reloaded, on a truck this time, and hauled to Sioux City, Iowa, where they're slaughtered. Then the meat is loaded onto refrigerated trucks for the long travel to Oakland, California, to the Niman Ranch Pork Company.

Not many farmers produce organic pork, not even for Niman Ranch. Bill Niman and Orville Schell, who would become dean of the Graduate

School of Journalism at the University of California, founded the company in the 1970s with a modest aim: producing fine-tasting beef, pork, and lamb. The company slaughters between eight hundred and a thousand hogs a week purchased from growers who agree to stipulations. The pigs are fattened in pastures, unlike the hogs packed by the hundreds and even the thousands in corporate hog factories known as confined animal feeding operations. Growers must not use antibiotics or growth hormones.

Of those thousand or so hogs butchered by Niman every week, only five to ten are categorized as organically grown. Those are Paul Gebhart's Illinois pigs.

The criteria for organically grown hogs remain murky, which Niman general manager Rob Hurlbut blames on the Agriculture Department's foot-dragging in adopting rules for the government's new organic certification. Most important is what they eat and don't eat. Organic animals can't be fed grain that is genetically modified, a problem because commodity soybean and corn are blended together from different fields.

For Paul Gebhart's pigs, that means dining on nongenetically engineered, tofu-quality soybeans that he grows himself.

"We're talking about the finest soybeans that we sell to the Japanese for fourteen dollars a bushel," Paul says. "I'm feeding my hogs the same beans that we feed the Japanese. They say the digestive tracts in pigs and humans are a lot alike."

ROAD TO NEW CITY

When I heard about Paul's organic operation, I assumed it would be a snap to find. At age twenty-two, I had lived in a farmhouse in New City, Illinois, which was not a city at all but a crossroads with a bait shop and a decrepit grocery. New City loafs along the Sangamon River on the way to Edinburg, the town nearest to Paul's farm.

Maybe because I'd lived in his part of the world, Paul stinted on directions. Or maybe I'm not as back-road smart as I thought. Talking to Paul on the phone, I wrote: "Near New City Road, look for old filling station. Go straight east out of town to stop sign. See dirt road. Keep

Organic farmer Paul Gebhart's pork travels from Illinois
to one of the finest restaurants on the West Coast.
Like many organic farmers, he worries about "genetic
pollution" from neighboring farms where modified crops
are planted.

going east to blacktop. Look for grain bins. Go north to stop sign. Keep
going to dirt road. If muddy go one mile east and one-half mile north.
Look for windmill."

At any one time on the Illinois flatlands, I could see five sets of grain
bins and a couple of windmills. When I telephoned Paul from the road,
I was miles away from his farm. He gave me a new set of directions just
as confusing. I finally guessed my way up a lane surrounded by weeds,
where a man in his late forties with a beard, bright blue eyes, and a faded
blue T-shirt was waiting.

I was told that Paul would be good to meet not just because of his organic pigs but for the outspoken views about genetic engineering he spread around the countryside. About food, too. As we walked up his lane, he would snatch a handful of weeds, stuff them in his mouth, and chew. Some of the same plants that fall victim to Roundup on other farms I had visited were, to Paul, culinary delights. Particularly lamb's-quarter, which he insists is better than lettuce and more nutritious. I ate some. I would call it a vigorous chew.

"You bring some sheep here, and it's the first thing they eat; it's how it got its name," Paul says. "People don't know what to eat anymore to get what their bodies need. Take fat people; they eat potato chips, and their body is telling them that they need to eat forty pounds of them for nutrition. Pretty soon they're as big as that door. The animals that walk on two feet are dumber than the animals on the farm."

Paul and Ruby Gebhart converted to an all-organic operation in the mid-1990s. It took them that long to get "clean," Paul said, sounding like my friend in a 12-step program. They own three hundred acres of good soil, probably a million dollars' worth of ground. But two-thirds of it remains un-cultivated grass. Paul's philosophy is antithetical to Earl Butz's famous ad-monition to American farmers to "plant fence row to fence row."

Besides pigs, Paul and Ruby sell organically grown chickens and tur-keys. The birds, like the swine, live and dine in grassy fields. Hiking back to a pasture, I see the chickens in huge cages on wheels that Paul scoots every day so that they rest on top of fresh greenery. Both the chickens and the turkeys sell for $1.50 a pound, and Paul says he's making a fantastic return. His pork sells for about 50 percent higher than supermarket prices.

People drive a hundred miles up from St. Louis and nearly that far over from Champaign-Urbana, where the University of Illinois is situated, to fill their coolers with Paul's meat. In his farmhouse he shows me some of their letters. "Thank you for producing safe food in a chemical world," a woman had written.

"I can put a better chicken in your Crock-Pot," Paul tells me. "It's like Dizzy Dean said," he adds, alluding to an old Cardinals baseball pitcher famous in these parts. " 'If you can do it, it ain't braggin.' "

BLOWIN' IN THE WIND

Paul will talk pigs till the cows come home, but what really gets him going is genetically modified seed. I happen to have a bag of them with me, and Paul says, only half-jokingly, "Don't spill any of 'em around here."

His fear seems overblown considering that I could kick a rock from his drive into a vast field of Roundup Ready soybeans across the road. It is a field that troubles Paul, but not as painfully as the gene-altered corn bordering his farm. He speaks of a fear that I have been hearing often from many organic farmers, what they call "genetic trespass."

Paul and others worry that they will be unable to maintain the integrity of their organic operations with pollen from genetically altered corn blowing in the wind. It is not an unreasonable fear: Tortilla-makers had begun rejecting corn with modified ingredients, and they would become more vigilant in the coming months after a widely publicized recall of taco shells grown with Bt corn suitable for animals only.

Genetic contamination is a vexing new side issue of the technology that promises to make work for farmers easier and more profitable. What levels of unnatural DNA will be allowable if food is to remain organic or "identity preserved"? And who will be liable if organic growers get their food rejected? These are real and troubling issues that the government and the agriculture industry have only begun to consider. Paul has been thinking about them, and asking discomfiting questions.

Five months earlier, Paul telephoned his insurance agent. "I hate to bring this up," he began. "But I have your farm insurance and liability. What I want to know is, when I get ready to sue my neighbor for contaminating my crop with GMO—my neighbor that I know you also insure—will you stand by me or will you stand by him?"

After several seconds of silence, the agent replied, "That's a good question. I'll have to find out about that."

Shortly before my visit, Paul ran into his insurance man in town. "I still haven't got your answer," he was told.

con #2 –
farmer p.152

farmer
con
#2

e," Paul continues, "they're saying that if you have
MO, you can never be certified organic. The GMO
be in the soil. Just like the DNA of the woolly mam-
buried in icebergs. That seems a little drastic to me. But
we're dealing with, if this is where we're headed, we have
som... s here."

When Paul talks about genetically modified crops, it seldom has anything
to do with efficiency or yield or any of the measures of farmer success. Paul
has never planted an engineered seed and he doesn't know firsthand their
value to farmers. What Paul talks about is corporate control of farmers and
their land, and the fears he conjures may seem conspiratorial, even wacky.
But to farmers like Paul they are real, and as the debate over GMOs grows
in the United States, such ideas are likely to be aired more fully.

"It's like the Eskimo and the wolf," Paul says, speaking of the allure
of the technology. "The Eskimo freezes the handle of his knife, blade up,
in the ice, and he puts seal blood on the blade. And then the wolf comes
along and licks the seal blood, but cuts his own tongue. Then the wolf
starts bleeding and he tastes more blood and he licks harder. Finally, the
wolf bleeds to death."

What Paul is prophesying is a sequential loss of control by farmers over
their land, perhaps even the land itself. In recent years, farmers have
watched the new order take shape around them as chemical and seed
companies merged. To get the benefits of the new genetic technologies,
farmers are signing contracts that give these new entities the right to mon-
itor what is grown. Multinational grain giants like Cargill also have gotten
into the business of lending, which poses a new threat.

evidence

I tell Paul that something is missing in this picture of farmers losing
their land. He says, "All the companies would need to do to eliminate
the farmer is to put in a gene that makes the crops perennial rather than
annual. That may be twenty years away, but it's not unthinkable at all.
Pretty soon, they'll say, 'Paul, we don't need you anymore. You can just
go into town and get a job. If you want to run the combine, you can.
But we can just go hire somebody off the street to do that.'

"They'd have you by the short hairs. It would be the end of the existence of the farmer as we know them today *and* corporate control of the food supply. And if you control the food supply, you control the people. Gene modification is the means to the end of the farmer. It might seem far-fetched, but you can put my name on it. I just hope my neighbors aren't the wolf licking the blade."

Listening to Paul, I hear something else that is worrying the big boys more than the spread of Paul's theories. He's making money, like everybody else in the burgeoning organic industry. And he expects uncertainties about GMOs—even if they are unfounded—to make him a lot more.

He fishes a chart from a desk drawer. It shows farmers' share of profit from what they grow plunging during the twentieth century. The same chart shows the profits shooting upward for companies selling machinery and chemicals and for companies selling what the farmers grow.

What I see sitting across from me at a kitchen table is a businessman with a plan.

"I'm trying to eliminate these guys and steal the market from these guys," Paul says. "My nature is to squeeze a dime so hard that three nickels jump out of my hand. I only buy two things new, gum and toilet paper. And I don't chew gum."

Paul's blue eyes are dancing now, and he speaks excitedly. "What will it take to break corporate agriculture? You only need to steal a small part of their big pie, and they're done."

GOING AND COMING

Leaving Paul's farm, I take the scenic route through New City, past the farm where I used to live. Replaying Paul's words, I'm musing about whether war will break out in America between the establishment food industry and the insurgent organic trade. I'm wondering whether the organic folks are as pure of motive as they carry on and whether the big companies might try to flatten them.

I wind along the Sangamon River up to where the land spreads out, and I'm getting excited. It's been over twenty years since I've been back to my old farm. Maybe I'll knock on the farmhouse door. When I reach the New City intersection, the bait shop and the old grocery are gone. At the corner is a windowless white building that looks like a shed. A sign in front of it says Church of the Coming Lord. Painted in black on the side is a scary-looking Jesus.

I pass the neighboring farmhouse where I kicked in the door after the farmer who lived there shot my dog—for chasing his sheep, he said. Another half mile and I'm almost home but . . . something is different. My farmhouse, the stone garage, the shed, and the barn are gone. No longer do they exist. I wonder if I'm in the wrong place, but I see the old road leading back to the river.

I get out of the car and walk to where the farmstead once sprouted, and I see a sign that tells me where I'm standing: in the middle of a field of genetically modified soybeans.

DINNER

With a few phone calls, I learn that Paul's pork ends up in at a place few Illinois farmers get to visit: Chez Panisse, the restaurant in Berkeley made famous by Alice Waters, who, among her awards, has been called the best chef in America and one of the ten best in the world. Waters named her restaurant after a character in Marcel Pagnol's 1930s' trilogy of movies, *Marius, Fanny,* and *Cesar,* and she calls it "an homage to the sentiment, comedy, and informality of these classic films."

Waters has clung to the principle "fresh, local, seasonal" since Chez Panisse began offering its fixed-course fare in 1971. I'd eaten there on my newspaper's dollar a couple of times covering political campaigns, and I recalled that no one at a table of reporters was foolish enough to grab for the check.

I was eager to tell Paul what I had found out, and over the phone I read him the dish of that day, a dish that he could take credit for. I also told him what came along with it:

Autumn vegetable antipasto

Brodetto *of local rockfish with spinach, garlic croutons and red pepper mayonnaise*

Spit-roasted Niman Ranch pork loin with lima beans, Romano beans, escarole and chanterelles

Gelato di ricotta *and espresso granita with chocolate sauce*

I also mentioned the price: sixty-five dollars.

"That's one plate?" Paul asked.

When I told him it was, along with a 15 percent service charge, 8.25 percent sales tax and an eighteen-dollar corkage fee, Paul was silent for a moment. When I asked him what he thought about the final destination of his pigs, he replied, "It's really quite an honor; I'm really proud of it, and I'm gonna tell people."

I could hear his wheels turning before he added, "I'm gonna have to figure out how to get sixty-five bucks a plate for the food I'm selling."

11

GENES OF THE JUNGLE

*Since wealth and technology are as concentrated in the North as
biodiversity and poverty are in the South, the question of equity is
particularly hard to answer.*
—Jonathan Lash
President, World Resources Institute

*One of these days I'm going to arm my boys so we can get rid
of these shitty gringos.*
—Gabriel Garcia Marquez;
Col. Aureliano Buendia in *100 Years of Solitude*

As JONATHAN GONZALES strips leaves, a howler monkey screams in
protest. After a three-hour hunt in a Panamanian jungle, the young bioprospector has discovered a worthy specimen: tender shoots from a mysterious vine curling along the forest floor. After he has gathered a pile of
the leaves, he seals them in a plastic bag, which he in turn places in a pol-
ystyrene box.

"The chances are low to hit the jackpot, but who knows?" Gonzales
tells me.

Gonzales, a Panamanian employee of the Smithsonian Institution, is a
collector in a bioprospecting program sanctioned by the United States
government. In the rain forest where he hunts, on Barro Colorado Island
near the Panama Canal, twelve-foot-long boa constrictors slither and furry
agoutis stroll in search of lunch. Here are "suicide" trees, *Tachigali versicolor*,

that die when they finally flower, and "beer belly" trees, *Pseudobombax septenatum,* with trunks resembling the torsos of several of my neighbors. By late in the morning, it feels saunalike inside my long-sleeved shirt and rubber boots borrowed as a barrier to the bullet ants.

The rain forests in Panama are rich with a diversity of species second in the world only to Costa Rica in concentration. Barro Colorado Island alone has more species of plants than the European continent. That is why bioprospectors are arriving here, and why the tropics are becoming a frontier in the Genetic Age.

Like the Gold Rush of the mid-nineteenth century, bioprospecting—the mining of genetic resources—rushed forward on the wings of scientific discovery. With new techniques derived from DNA manipulation, plants can be tested for their valuable properties hundreds of times faster than a decade ago. Now that the power exists to transfer genes across species and even kingdoms, the tropics offer not just the makings of medicines but also raw materials for designer crops and foods.

It is a race; global rain forests are shrinking at the rate of one hundred acres a minute. By the year 2020, another 10 percent of the world's species of plants, animals, and microorganisms will become extinct as a result of deforestation, sprawling growth, and unsustainable practices in farming and fishing. When a species disappears, so does its unique DNA—along with all the valuable properties it may hold.

It is a race with many hurdles, among them new restrictions on access imposed by countries weary of exploitations of their genetic resources. By their reckoning, the pursuits of the bioprospectors can rightly be called "bio-piracy."

Medicines for many ailments, even cancer, have been discovered in the tropics. For centuries, plants have been hunted, traded, and plundered. Now, because of genetic engineering, the potential of the world's rain forests is greater than ever. We know little about the bioprospecting trade because companies keep their expeditions quiet. Never have they wished to draw attention to the hunt for substances worth millions of times their weight in gold. Since a backlash in the early 1990s, their trade has grown even more controlled. The biodiversity Mecca, Costa Rica has passed a law controlling access—as have plant-rich Ecuador,

Peru, Venezuela, Bolivia, Colombia, Ecuador, and the Philippines, among others.

An exception to the rule of secrecy is the United States government, which sponsors private collectors and oversees sensitive negotiations. I've tracked down several of these arrangements, and now I have journeyed to Latin America, stopping first in Panama, to visit government-sanctioned expeditions.

HAND IN HAND INTO THE JUNGLE

Government partnerships were meant to assuage critics while avoiding the ethical transgressions of the past. Egregious exploitation over the years prompted the adoption of the International Convention on Biological Diversity at the 1992 Earth Summit in Rio de Janeiro. The treaty-like convention has three aims: conserving biological diversity; promoting sustainable use of natural resources; and equitably dividing benefits from the use of genetic resources. Nine years after the landmark agreement was approved, 180 countries had ratified the agreement. The first was Mauritius; the 180th Zimbabwe. There remained one holdout of note: the United States, which had signed the convention but not ratified it, primarily because of the objections of the Republican-controlled Senate to participating in another global organization.

Outside the convention, the American government has invested more than $20 million to foster individual bioprospecting agreements, called International Cooperative Biodiversity Groups, which join the sponsoring companies and American researchers with the institutions and tribes of the tropical lands where most of the world's genetic resources are found. Skeptics view this United States involvement as a form of corporate welfare given those that stand to benefit, a list that includes Bristol-Myers Squibb, DowElanco, Monsanto, Glaxo Wellcome, which would become Glaxo-SmithKline.

Further complicating the bioprospecting business is the U.S. system of patenting. Patents are the bedrock of business innovation, but many in developing lands just don't see how companies can claim to own genetic material and living beings. Indigenous people have an especially hard time grasping how knowledge they've used for centuries can suddenly become the intellectual property of a foreigner thousands of miles away. Plant-patenting has become a flash point in North–South relations. In India, tens of thousands of people took to the streets in 1998 to protest a patent awarded in the United States to govern the marketing of varieties of India's revered basmati rice grown in the Western Hemisphere. Another patent, awarded to the University of Mississippi on turmeric, was canceled after challenges from India's government.

Jonathan Gonzales has no ethical problem with patents from bioprospecting. Given the furor about exploiting the locals, it's not hard to figure out why I've been assigned to a Panamanian collector rather than a corporate scientist. It's fine with me because Jonathan is an amiable young man in his mid-twenties and full of information. (In this forest there are 1,369 plant species and about 1200 howler monkeys who gambol about, nineteen to a troop, he tells me.) He is also not given to obfuscation which, I have observed, seems to be a genetic trait among bioprospectors.

What Jonathan has plucked today from a swath of forest floor opened to sunlight by a fallen tree will be assayed for its capacity to combat dengue fever, malaria, and HIV. Other samples taken in this project will be tried against crop pests.

"It is good for Panama not only because of the royalties we would get, but because of the new opportunities to train scientists and the acquisition of sophisticated equipment," he tells me, speaking of the new arrangement. "We are all happy so far."

But I know from talking to others that not everyone locally is so pleased.

After spending the morning in the forest, we return to camp for lunch. This is the routine of bioprospectors in the tropics: Jungle-traipsing is conducted in the morning, before the heat becomes all-consuming and the afternoon rains arrive. We lunch at the Smithsonian Tropical Research Center, which has been here since the arrival of scientists from the United States in 1923 to control the disease-carrying mosquitoes impeding construction of the Panama Canal. I am momentarily troubled when I see a sign near the dining hall that reads: "Please turn in dead monkeys to Mr.

Boniface's office and receive payment." I ask no questions but turn down the cook's offer of meat.

For the bioprospector and his visitor, the next stop is the laboratory. It takes about twenty grams of chosen vegetation, half to be tested and the other half to be labeled and frozen in the event something promising shows up in the tests. With a razor blade, Jonathan chops up the nameless leaves and then pulverizes them in a kitchen blender, as if he were making pesto. Then he mixes the green confetti with methanol in order to purify his extraction. When he's done, he has a slurry that resembles pea soup.

The extracts from the search are tested at the University of Panama in Panama City for disease-fighting properties. There, scientists determine whether the extracts from the jungle plants will kill live cancer cells and other disease lines that are kept in rusting refrigerators. That is just the beginning of a protracted process, and it may take ten thousand assays to get one positive "hit"—which is the ticket to proceed with further testing and the means by which you mollify those who are paying for your jungle treks.

The initial assay for disease-fighting powers is straightforward. You combine the purified extract with the live cells growing in a medium, then peer through a microscope to see if the cells are succumbing. If that is the case, you have what researchers call biological activity, and that's when the shroud of secrecy usually falls.

Assays for natural resistance against white flies—which would amount to a golden discovery in the crop sciences realm of genetic engineering—are proving to be more difficult. It seems that the researchers are having trouble getting the extracts inside the white flies, or close enough to them, to determine if they might have the makings of a pesticide. Todd Capson, a Smithsonian scientist and participant in the bioprospecting project, put it this way: "These guys are having one hell of a problem. What do you do if a white fly turns up its little nose and says, 'I don't want to eat this'? It's not nearly as easy as working with cancer cells growing in a medium. You put your stuff in there and see if it kills it. But the white flies, they're jumping all over the place. Fortunately, cancer cells can't fly. If they did, we'd all be in trouble."

A scientist at the university, Mahabir Gupta, is thrilled about what the project can mean for his lab and for Panama. He also says he understands why indigenous people are wary about the arrival of the bioprospectors. "They think they might have the keys to a gold mine, and they don't want to give them away," he says.

Nearby at the Gorgas Institute of Health Research in Panama City, the extracts are tested further to see if they possess properties to fight malaria, dengue fever, and parasites—the bloodsuckers and body invaders that kill seventy thousand people yearly in Latin America alone. The Gorgas Institute is named for William Crawford Gorgas, who in the 1920s was in charge of quelling the yellow fever that killed scores of workers on the Panama Canal. By all accounts, the once-famous institute has lost its luster, a process accelerated in 1989, when the United States invaded Panama and Congress cut off the institute's money. When I stopped in to see Dr. Eduardo Ortega, the director, he told me that bioprospecting could be the means to restoring Gorgas's respectability and recruiting Panama's next generation of scientists.

"For us, this is like something from the sky that has come to help," he said.

"DIRTY BUSINESS"

The new bioprospectors bring equipment, cash, and hope. It may take plenty of each to outweigh the sins of the past.

Everywhere the bioprospectors land, natives are suspicious, and for good reason. Before heading to Latin America, I tracked down experts who had witnessed these sins. "Companies and their agents would come in and offer $50 or $100 and take what they wanted, and that would be that," said Jerrold Meinwald, a Cornell University chemist who directed an International Cooperative Biodiversity Group hunt in Costa Rica.

"It was a dirty business," echoed ethnobotanist Brent Berlin, of the University of Georgia. "There was basically no recognition of property rights or traditional uses of medicine." Berlin soon discovered that the indigenous people hadn't forgotten the seamy dealings of old when he took the leadership of another Biodiversity Group, which was seeking medicinal plants in the highlands of Chiapas, Mexico. Traditional Mayan healers, unimpressed with promises of shared royalties, pressured the Mexican government to stop Berlin and his corporate partner, Molecular Nature Limited, of the United Kingdom, from their work.

In the 1992 film *Medicine Man,* Sean Connery finds not only cancer-fighting plants but also romance in the jungle. In reality, bioprospecting is a perilous pursuit with disappointment more often the rule. Bioprospectors pack antidotes for the bites of deadly pit vipers. Near the border dividing Peru and Argentina, they sidestep land mines planted by soldiers. They bring back to their civilization parasites that aren't specimens.

Success is elusive. "There are people who think there's green gold out there," Gordon Cragg, of the National Cancer Institute, told me. "But the chances aren't very good."

Nevertheless, if they're lucky, a trip into the jungle can make their companies, their institutions, or the bioprospectors themselves rich. They might even stumble upon a discovery that alters the equation in the fight against disease or, these days, discover a protein or an enzyme that will be genetically engineered into food. Two-thirds of cancer drugs and many of the world's best-selling pharmaceuticals are derived from natural products. In 1997, seven of the world's twenty-five top-selling drugs, with a combined sales of over $11 billion, were derived from natural products. And much of this drug supply originated in the tropical belt that spans the globe, a biodiversity wonderland that sprouts two-thirds of the world's 250,000 plants—which includes the vascular plants, such as seed plants and ferns, and the bryophytes, such as mosses and hornworts. (There are an estimated 25,000 undiscovered species of plants, most of them in the tropics.)

Two cancer drugs, vincristine and vinblastine, developed by Eli Lilly & Company in the middle of the twentieth century from the rosy periwinkle of eastern Madagascar, have yielded hundreds of millions of dollars in profits. But Madagascar doesn't share in those proceeds. Nor is the island nation an exception: There are no known cases of countries or indigenous people receiving royalties from products derived from their lands. That's why plant-rich countries have clamped down and why bioprospectors have been forced to operate in a new way. The National Cancer Institute, known as "Uncle Sam's Drug Company," has seen its own bioprospecting curtailed around the world.

These days, bioprospectors hope to discover more than just drugs. Over thousands of years, many plant species have developed chemical compounds that protect them against insects, fungi, and viruses. The same potent chemicals extracted for medicines might be engineered into plants. Beyond crop protection, the researchers want genetic materials that could yield better nutrition in food and benefits even more profound. David Corley, a natural-products chemist at Monsanto, offered an example of the potential when I visited him in St. Louis. If researchers find Amazonian lands where prostate cancer is rare, he said, they would try to learn if Indian men are protected by something they eat. If they can identify a biologically active agent that girds men against prostate cancer, they might be able to engineer it into foods.

In late 2000, six International Cooperative Biodiversity Group projects were operating in ten countries in Latin America, Africa, and Asia. Three of these projects—in Panama, in Madagascar, and in Chile and Argentina—included the search for genetic materials that could be engineered into plants. In the South American project, the bioprospectors hunting in arid lands hoped to discover antifungal agents. Monsanto's Corley summed up the promise driving industries into the bioprospecting game. "There are breakthrough technologies coming along that are just going to blow this whole thing open," he said.

"VINE OF THE SOUL"

No barrier looms larger to bioprospectors than their fractured relations with indigenous people. In the Philippines in the mid-1990s, the Talaandig tribe seized a collection in the works and accused bioprospectors of having "gravely transgressed our sacred customs and traditions."

In Ecuador, a California entrepreneur, Loren Miller, was declared "an enemy of the indigenous people" and threatened after obtaining a patent in the United States on ayahuasca, a sacred hallucinogen whose name translates to "vine of the soul." Ayahuasca is prepared from the bark of a nondescript bushy tree I found growing in an Ecuadorian jungle. It is a drug not for the patient but for the doctor; shamans drink it to journey in the spirit world, where sickness and death—and the knowledge to overcome those ills—are believed to originate.

Miller, who is a pharmacologist by training, set up in California a company called International Plant Medicine after being given an ayahuasca plant in the 1970s and further cultivating it in Hawaii. His aim was not to profit from the plant's hallucinogenic properties, even though some have taken that route by means of Internet sales. Rather, he wanted to explore its potential in psychotherapy or perhaps as a cancer drug, he told me.

Miller's patent became a *cause célèbre* among indigenous groups, particularly the Coordinating Body of Indigenous Organizations of the Amazon Basin, whose leaders declared him an enemy and announced that if he or his associates return to their lands, they "will not be responsible for the consequences to their physical safety." Tribal leaders adorned in shell necklaces and exotic bird feathers showed up in Washington in the spring of 1999, trying to persuade the United States Patent and Trademark Office to revoke Miller's patent. "Our ancestors learned the knowledge of this medicine, and we are the owners of this knowledge," Antonio Jacanamijoy, who headed the indigenous council, told me.

In my conversations with Miller, he said that his patent remained unpursued, lying in a drawer, but that he didn't intend to abandon it without a fight. Indeed, he convinced the Inter-American Foundation, a government agency, to cut off aid to the tribal council after having given it more than $500,000 annually in recent years. "I'm fighting for my life," Miller said. "What I'm trying to do is make it safe for me and the tribe that gave it to me to go back and to grow the plant."

Brent Berlin was traveling in Peru in the 1980s, when members of the Aguaruna tribe sank his dugout canoe, sending a year's worth of botanical samples swirling down a muddy river. It was the same fierce tribe, he told me, believed to have burned the jungle camp of German director Werner Herzog while he was filming the movie *Fitzcarraldo,* the story of a nineteenth-century Irish entrepreneur bent on harvesting rubber from the Amazon jungle.

Ethnobotanist Walter Lewis, of Washington University in St. Louis, discovered that the Aguarunas' attitude toward outsiders had mellowed to the extent that they were slower to turn to violence. Lewis had begun bioprospecting in Peru for valuable jungle plants in one of the early government-backed Biodiversity Group arrangements when the Aguarunas concluded they didn't like the deal that had been struck. They demanded return of three hundred specimens, kept $10,000 of Monsanto's money, and declared that they didn't want to see St. Louisans in their jungles anytime soon.

Lewis contends that his problems had a lot to do with Evaristo Nunkuag, an Aguaruna leader who has won international awards for helping his people and the rain forest. His reputation is such that he is known in the jungles simply as Evaristo. He and Walter didn't remain on a first-name basis for long. "It was a horrible experience that I wouldn't want to go through again," Lewis said when I visited him in St. Louis.

Indigenous tribes who have been fending off multinational companies might say the same thing. Lewis's initial negotiations with the Aguarunas provided a 1990s' lesson in how not to deal with people who have grown protective over their genetic resources, his critics say.

Lewis, who was in his late sixties when I met him, is widely known in his field. He has written books on bioprospecting, among them *Medical Botany: Plants Affecting Human Health,* which he coauthored with his wife, Memory Elvin-Lewis. In one of his papers, he described his quest—and his profession. "This is what a handful of researchers do and do well, by obtaining ethnobotanical information from intact indigenous peoples. But today it is scarce and growing rarer."

With $2.2 million in government funding, Lewis set out in 1994 in search of healing secrets from Aguarunas, a tribe that has worn the label "headhunter." Soon, his problems would halt his jungle pursuit in its tracks, with the Aguarunas complaining to the National Institutes of Health in Washington that they were victims of "manipulations and aggressions."

"People felt they were being taken for a ride," said Joshua Rosenthal, of the National Institutes of Health, who was sent to investigate. According to various accounts, the $10,000 royalty was spent on health care and education for the tribe.

Lewis insists that the deal was fair for the times. A few years afterward, tribes are demanding—and getting—promises of higher royalties and up-front payments. Lewis acknowledges his blunders. "I think the words 'political naïveté' were used to describe me, and I couldn't agree more," he told me.

Lewis succeeded in forging a new arrangement with another group of Aguarunas and, before long, with the Indians' blessing, his team was back in the Peruvian jungle. I saw the new attitude of collaboration when I visited Lewis's new partners in Lima. They may have hunted heads in the not-too-distant past, but now they carry cell phones. And they had learned the capitalist way of driving a hard bargain. Indeed, they threatened to walk away from the negotiating table.

"We said that our plants aren't waiting for anybody in particular," Cesar Sarasara, the president of an Aguaruna federation, said when recounting how the deal came together.

The tribe also was sufficiently astute to demand something unprecedented in the bioprospecting trade: a "know-how" license, in which Monsanto, the corporate partner, agreed to pay them not just for their plants but for their knowledge. That covers tea, paste, slurry, and the Aguarunas' secrets for cooking, boiling, and distilling, according to a copy of the agreement the Indians showed me. When I visited them, the Aguarunas were getting more than $30,000 a year for allowing the bioprospectors in their territory. If a promising product gets to clinical trials, they would receive a six-figure payment. And if a successful product results, the Indians will share royalties that could make their federation wealthy.

As the diminutive but powerfully built Sarasara put it, somebody always arrives to take something from the jungle, so the people who live there must be ready to bargain. "Tomorrow, we may be talking about our gold, or our oil, not our plants. We Aguaruna people have to protect our biodiversity, but we also have to take into consideration the reality of the moment. For now, this is an arrangement of hope," he said.

At the Jatun Sacha jungle preserve in Ecuador, botanist
David Neill has felt the heat from critics of bioprospecting
despite his labors on behalf of conservation.

"BLOOD OF THE DRAGON"

The travails of the business are enough to make a bioprospector lie low.
I discovered as much after I crossed the Andes to track down David Neill
in Jatun Sacha, an Ecuadorian jungle preserve. In the rain forest, I watched
as Neill dug his knife into a white-barked tree and stepped back to observe
the consequence. Out of the gouge oozed a blood-red substance that he
hopes will enable the Waoranis, Quichuas, and other nearby Ecuadorian
tribes to cash in.

For centuries, indigenous people in the upper Amazonian Basin have relied on the red drip from the tree they call *sangre de drago*—blood of the dragon—to treat wounds and ailments. Neill is working on a plan for indigenous people in South America to grow the trees, which may one day become valuable as an antidiarrhea medicine if the Food and Drug Administration overcomes its wariness.

Neill believes that by growing stands of these medicine trees, Indians can develop the wherewithal to avoid slashing deeper into rain forests to carve out subsistence croplands. "With a steady income, they won't need to keep cutting more forests and moving on," he says.

The bearded Neill, who is in his mid-forties, bears a resemblance to Abraham Lincoln in the president's pre-Washington years, when he practiced law in Springfield, Illinois. He is a staff member of the Missouri Botanical Garden, whose researchers conduct far-flung missions around the world. MoBot, as it is known back in St. Louis, where the Garden spreads out over seventy-nine verdant acres, is one of three institutions that collects plants for the National Cancer Institute. The Garden also has collected for Monsanto, and more recently, in an International Cooperative Biodiversity Group arrangement, its botanists have been dispatched to Madagascar and Surinam.

The Missouri Botanical Garden, known widely for plant research and promoting the preservation of biodiversity, has endured criticism for its alliances with Monsanto. Similarly, Neill has been demonized for his connections to bioprospecting.

David Neill would seem an unlikely candidate for vilification. He is known for engineering conservation projects; he founded Ecuador's national Herbarium, which catalogs and preserves plants. And he cofounded Jatun Sacha, (pronounced hot-TUNE SAHTCH-uh), a five-thousand-acre reserve in the upper Amazonian Basin that has more species of plants than the United States and Canada put together and is dedicated to conservation, research, and teaching. The rock group Grateful Dead thought enough of Jatun Sacha to give it $10,000. The International Children's Rain Forest Network raised money so that Jatun Sacha could expand its holdings.

But Neill has felt the backlash from indigenous groups over the ayahuasca patent and other perceived transgressions. "I just about got blown out of this country," he told me one evening as we drank beer in a thatched hut in the jungle. "I was criticized very, very hard. I decided to let somebody else go through the regulations. I decided to figure out another way to make botany pay."

Neill's botanical specialty is beans. He's a bean counter in another sense, too: He raises money for his conservation projects and tries to turn a profit, when he can, from a scientific pursuit that is growing obsolete with so much of the world plundered and discovered. In the past, he has engaged in bioprospecting, and he has been hired by oil companies—ARCO, Triton, and Petro-Canada, among others—to conduct environmental studies in advance of drilling.

Neill was preparing to take part in another bioprospecting project with a New York–based company. But it was not to be. With tensions running high, Ecuadorian officials refused to give the go-ahead; indeed, none of the five Andean Pact nations had consented to a new project in more than two years at the time of my visit. Neill was shying away from new bioprospecting, waiting for the air to clear. "It's too political, too highly charged. Twenty years ago, nobody asked questions about it. Now, people are much more aware of their rights," he said.

After walking in the jungle the next afternoon, I retreated to my tiny cabin—which has its name, *Huanduj,* carved into a wooden shingle outside—to examine welts and nibbles from insects. *Huanduj* is a hallucinogenic plant, characterized by its huge, bell flowers, and I resolved to find one of the bushes, in hopes of expanding my consciousness with regard to bioprospecting. I reconsidered when I found a soggy reference manual and read about the plant: "The dosage is very, very critical. . . . You can go crazy," it said.

I did not need to risk a Kurtz-like madness in this jungle, I concluded. I found myself sufficiently unsettled when the rains arrived. It was a ferocious, lashing rain that pounded my cabin. Drinking two-dollar rum, waiting for it to stop, I wrote in a notebook: "After a six-hour drive over the Andes, much of it unpaved, I have arrived in search of the same bounty that is attracting companies and their emissaries: knowledge of genetic resources. Only a few years ago, these plants were regarded as the common heritage of humankind. Now, they are a commodified entity of the Genetic Era, sooner or later to be auctioned to the highest bidder."

In the morning, I decided unexpectedly to return to Panama to visit the Kuna, the indigenous people there. Before riding back across the Andes, often high above cloud forests, in a rainbow-colored bus that would have appealed to the psychedelic yearnings of Ken Kesey and his Merry Pranksters in the 1960s, I sought out David Neill once more. We hiked to a grove of 650 *sangre de drago* trees, and I brought blood from one of them with my own knife.

Aware of my broader interest in genetically modified food, Neill told me that he was concerned about the potential safety threats from reordering the genetic materials of crops. I found that assessment reassuring coming from a scientist who understands plants and whom I respect. But, Neill added, what worries botanists is the trend, accompanying biotech and intensive farming, of raising fewer and fewer crop varieties. "You run the risk of seeing whole genomes wiped out by something that comes along in nature. And we never know when that's going to happen. Nature is always very chaotic," he says.

Neill reminded me that his pursuit these days is agroforestry—cultivating trees—not bioprospecting. For botanists like him, people who find joy in collecting and cataloging plants, biotechnology is more than a source of income; it is a connection to microbiology, the hot science of the day. "Science is changing," he said wistfully. "People like us are dinosaurs. We are the only people who read something two hundred years old. What we really want to do is so esoteric that nobody cares about it but us."

IN KUNA LAND

Traveling by motorized dugout canoe, after a bruising ride in a pickup, my journey to Kuna territory took nearly as long as crossing the Andes. Some thirty thousand Kuna reside in a narrow band of Panamanian jungle that stretches 120 miles along the Atlantic coast to the outlaw territory of drug cowboys and revolutionists along the Colombian border. Another twenty thousand or so Kuna live in Panama City and inland. Across Panama, you see Kuna women in their red-and-yellow head scarves, wraparound skirts, and reverse-appliqué *mola* blouses. On the streets they sell *molas,* cloths the size of place mats, adorned with turtles, fish, and fearsome spirits stitched in vibrant oranges, purples, and tropical hues.

If any tribe can handle the bioprospectors, it's the Kuna. They have fended off conquistadors, gold-hunters, turtle-raiders, and pirates the caliber of Henry Morgan. From their rugged jungle enclaves, many of them reachable still by plane or launch only, they have driven out pillagers, evicted squatters, and remained free from the reach of dictatorship. In 1925, it took the assistance of a warship, the U.S.S. *Cleveland,* but the Kuna triumphed in an armed rebellion against a new government's national guardsmen.

It is their tribal homeland, Comarca de San Blas, formally granted in 1938, that sustains the Kuna, giving them power. In a traditional village, where I have disembarked, the Kuna believe they have protection from a spirit realm that is every bit as frightful as a band of rapacious bioprospectors. Outside the perimeter of their village are *kalumars*—designated sanctuaries of spiritual plants, animals, and demons. In the hills and in whirlpools beneath the sea reside whole communities of spirits. Panamanian biologist Jorge Ventocilla coauthored a paper describing the transcendent importance of the land in Kuna cosmology:

Underlaid and animated by the realm of spirit, the Earth is a living being, the source of all life. It is the body of the Mother, who carries on her eternal task of regenerating the planet with newly born plants and animals from her womb deep in the bowels of the cosmos. The thick mantle of rainforest covering the mountains and valleys constitutes the Mother's green clothes. Simultaneously, the rivers are seen by the Kuna as the Mother's vagina, for their reproductive function, and as her breasts, because they nurture her creatures with her milk. She provides the Kuna with food, building materials, firewood and countless other materials they need to survive. Kuna leaders, in their nightly gathering sessions, constantly admonish their people to treat her and all of her things with respect.

Omar Torrijos, who ruled Panama for twelve years until 1981, coveted the Kuna land from his helicopter on his way to the Kuna-held coastline and told them so in 1979 during a meeting on the island of Nargana:

Why do you Kuna need so much land? You don't do anything with it. You don't use it. And if anyone else so much as cuts down a single tree, you shout and scream.

A Kuna leader, Rafael Harris, rose to speak:

If I go to Panama City and stand in front of a pharmacy and, because I need medicine, pick up a rock and break the window, you would take me away and put me in jail. For me, the forest is my pharmacy. If I have sores on my legs, I go to the forest and get the medicine I need to cure them. The forest is also a great refrigerator. It keeps the food I need fresh. If I need a peccary, I go to the forest with my rifle and—pow!—take out food for myself and my family. So we Kuna need the forest, and we use it and take much from it. But we can take what we need without having to destroy everything as your people do.

By no means are the Kuna backward forest-dwellers. With their Kuna General Congress and their villages, they have a highly defined political and social order. They are astute at business, inviting tourists to their islands to observe and photograph their culture, housing them in Kuna-owned hotels, and then dispatching them with a load of sometimes pricey *molas* and folk art. As one Panamanian analyst put it, the Kuna have treated Western culture like a supermarket, taking what fits them but rejecting most everything else.

The Kuna attitude toward bioprospecting is both wary and curious. They are wary about the environmental implications and the unnaturalness of genetic engineering, about which they are just learning. They are curious about the financial promise so far denied them. At the outset, they were not members of the International Cooperative Biodiversity Group, a deliberate decision that stemmed in part from a tenuous relationship with the Smithsonian. Organizers decided that the best way to handle the Kuna was to deal with them later. That decision followed an unfortunate dispute that had boiled over a few months before my visit.

In July of 1998, after twenty-one years of studying coral reef ecology, the Smithsonian was ordered to abandon its San Blas Research Station on Kuna land. Smithsonian researchers had studied reef ecology in the Atlantic Ocean off Kuna land to gauge the effects of climate change. The project at pristine San Blas had provided the baseline for studies throughout the Caribbean. No more. The Kuna accused the Smithsonian of high-handed dealings and theft of coral and other specimens from the ocean floor. The Indians were troubled by lights in the water at night, and rumors spread about a military operation in their midst.

"The scientists wouldn't tell people what was going on," Geodisio Castillo, who represents the Kuna organization that goes by the acronym PEMASKY, told me in his Panama City office. "Sometimes when you're Indian, people think you're ignorant so they don't talk to you."

The fracturing of the Kuna-Smithsonian relationship was a shame, a breakdown the Smithsonian has lamented as a failure in communications. The relationship had endured for nearly three-quarters of a century, since a visit by Kuna leaders to the National Museum of Natural History in Washington—the trip during which the Kuna plan to rebel against a new government took form.

If the goal of the United States–funded bioprospecting projects is to overcome tribal suspicions and mistrust of the past, organizers have a ways to travel with the Kuna. "They have no respect for us. They have their million-dollar projects and they don't even tell us what's going on," grumbled Jose Armando Palma, an officer in a Kuna political organization called ORKUM.

The International Cooperative Biodiversity Group bioprospecting project is bringing Panama equipment, potential defense against tropical diseases that the multinational drug companies have all but ignored, and hope for future rewards. "The world needs a success story, and everything is here," the Smithsonian's Todd Capson told me.

But it had offered nothing more to the Kuna than waterproof paper and lessons in collecting botanical specimens. That is troubling to Kuna like Horacio Rivera, who heads ORKUM, and who wonders how the people who inhabit 70 percent of Panama's rain forest can be left out. "We want to be part of this, but they won't let us," Rivera told me.

Deep in Kuna land, after my journey by dugout canoe, I found a similar attitude on the part of Rivera's father, Felicio, who is an Indian healer. It is an attitude suggesting that the owners of the world's treasure troves of genetic resources will share them. But not if there's so much as a whiff of greed and exploitation.

When fevers strike his village, Felicio, a tiny, earnest man in his early sixties, retires to the riverbank for medicine. He chants to the chosen plants, and then he carries their leaves back to his village in a basket woven from palm bark. The omnipresent malaria inflicts the Kuna, but the fevers come worse: Dengue fever lurks like a rain forest jungle cat, and villagers gripped by it know why it's called "breakbone disease."

Crouching outside his hut, the barefoot healer told me about some of his special trees and the bushes that Kuna have visited for generations. Sometimes, when I asked him their names and how he makes medicines, he shook his head. Before leaving, I did my best to explain bioprospecting, for which there is no simple word in his language. When I asked him if he would be willing to share his secrets with bioprospectors, he responded in a language that even indigenous people fearful of relinquishing their heritage seem to understand

"You bet," he said, without pausing. "But if I teach, they pay. Like school."

PLANTINGS SIX

The Roundup worked. My genetically engineered soybeans remained green and hearty, already bigger than my habanero peppers growing nearby. But the conventional beans I had squirted were a mess; within hours they had turned a sickly yellow and two days later dried to a brown.

Of course I realized later that I wasn't thinking like a scientist. I got carried away spraying and killed half of my conventional beans, and now I had just one row of them left. So much for yield comparison.

The weeds succumbed, and so did the poison ivy. I also administered a death squirt to a couple of Grandpa Ott's feisty morning glories, which had attempted to strangle my Roundup Ready beans. It says in my Roundup brochure that the chemical "is absorbed into the plant and transported through the plant, where it prevents the plant from making its own food. The plant starves to death, so when it dies, every part dies, including the deepest roots."

The Monsanto scientists who invented herbicide tolerance gave me more details. The Roundup blocks what's called the aromatic biosynthesis pathway, which leads to three amino acids. About twenty years ago—after Roundup was sold everywhere—they found which amino acid it was: tryptophane. They concluded that Roundup works during a step in tryptophane synthesis by blocking a particular enzyme. They weren't done yet.

They isolated the enzyme, purified it, and figured out its genetic code. That made it possible to clone it and engineer it back into petunias, which was a critical step in the development of herbicide tolerance.

Later, they found the perfect gene, and it didn't come from a cool, shiny laboratory with electron microscopes and people in white coats. They found it in the muck of a waste treatment pond behind a Roundup manufacturing plant.

For years, the microbe had sat there in the toxic ooze in a fenced-off plot toughening itself by eating Roundup, which it found nutritious. If there was anything on earth unafraid of Roundup, it was this bad-ass little microbe that ate it for break-fast, lunch, and dinner. They isolated this little fellow, replicated it millions of times, and that's what's used in engineered soy-beans today.

Among them are the flourishing genetically engineered plants in my backyard, which shook off their Roundup shower like prizefighters ridding their muscled bodies of sweat.

Part Three

BACKLASH

12

IN IRELAND, BEETS OF WRATH

There is not a single case of any damage having been done by
agricultural biotechnology to any person, any ecological system,
or to any environment.
—David McConnell: Geneticist, Trinity University, Dublin, 1998

I am troubled, I am dissatisfied, I am Irish.
—Marianne Moore, "Spenser's Ireland," 1941

IN THE LAND of the Great Potato Famine, rebellion is rising against the American company that reinvented the world's potatoes. Saboteurs have struck in County Carlow, slashing and trampling Monsanto Company's gene-altered test crop. To discover the source of such anger, I have traveled to Ireland. As I track down the Irish story—interviewing farmers, critics, Irish consumers, and company representatives—native-born resistance rises throughout Europe. Neither I nor anyone else knows where it will end; we have arrived in unmapped territory. Biotechnology may conquer Europe with barely a scrimmage. Then again, Europe may rise up to overthrow the upstart technology. Or something else may happen, something entirely unforeseen by the microbiologists who are fusing genes with the fervor with which the fathers of the nuclear age split atoms.

I'm betting that it will be consumers—not activists nor grocers nor life-science companies nor United States government trade negotiators—who determine the future. The people at the top of the genetically modified food chain think about science and farming and trade. The people at the bottom of the chain think about food, and that is why I am in Ireland, where food is a matter of the heart.

The Great Potato Famine remains a touchstone for the Irish, and its specter rises unbidden when people imagine tinkering with the genes of food. It was Ireland's worst disaster, scarring the collective psyche, much in the way the Civil War haunts Americans. From 1845 to 1848, 1.5 million of Ireland's eight million people died of starvation or from epidemics of typhoid and cholera that swept the land when potato crops failed four years running.

Potatoes weren't native to Ireland, but once they were imported from the New World—possibly by Virginia colonist Sir Walter Raleigh in the sixteenth century—they caught on. By the middle of the nineteenth century, potatoes made up the biggest part of the Irish diet. When the famine struck, another two million people left Ireland, most of them emigrating to Great Britain, the United States, and Canada. The failure of a single crop cut Ireland's population by over 40 percent.

Near Skibbereen in western County Cork, eight thousand people were buried in a single mass grave near the farm of Mairie and David Cregan. One wrenching day 150 years ago, a female ancestor of Mairie's carried her two dead daughters—one three years old and one three months—six miles over verdant hills to the mass grave. Coffins of that era were made of wood and constructed with a panel that opened. At the edge of the grave, it was tilted to let the body slid into the grave. The coffin then could be reused. But Marie's relative couldn't do it. Weakened by hunger, she nonetheless trudged the six miles back home with the bodies, resolving not to bury them until she could arrange for secure coffins to be built.

"People still have it in them. There is nobody around here who didn't have a relative who died," Mairie says.

The Cregans have six children spread over twelve years. Rory is the oldest, at sixteen, followed by Aoife, Siovhan, Aisling, Aidan, and Casey. Adding to their robustly large household are four teen-age foster children. The family milks thirty cows and grows vegetables: potatoes, carrots, broccoli, cabbage, and cucumbers. Strawberry patches produce abundantly, but the pails of newly picked fruit usually are emptied by the time the kids hike back to the three-bedroom stone farmhouse.

David, who was born in 1954 and who is eight years older than Mairie, has traveled the world working on oil rigs. At the moment, he is content living on the farm. When I ask David if genetic engineering can help him grow better food, he acknowledges that there might indeed be some benefits to herbicide-tolerant crops. But he has no intention of moving in that direction, he tells me, because of larger concerns.

"We don't need to produce more food," he says. Because of the pressures from the multinationals, he adds "we're concentrating on fewer and fewer crops already to feed the world."

David tells me he is troubled by what he calls a "dwindling sense of independence." But he is less upset at milk quotas and production controls from the European Union than by the perceived sins of multinational corporations. He is particularly irked at Monsanto, he says, because of what he is hearing about the technology fees the company slaps onto each bag of genetically modified seeds and the binding contracts with farmers. "They can tell you what to grow and where to grow it," he says.

In their newspaper and on the radio, the Cregans have learned about Monsanto's plan to test genetically modified sugar beets in Ireland. After Monsanto's first Irish test field was sabotaged, the company asked Ireland's Environmental Protection Agency for permits to plant its genetically modified sugar beets at ten more farms around the country. When protesters stormed test fields, the request was scaled back to five.

To Mairie Cregan it is unthinkable to tamper with the genetics of a food system that has grown abundant and reliable. As she serves cheese sandwiches to her children, she speaks to me in a tone that is more curious than it is fearful.

"It seemed like a space odyssey that was taking off from America, and the next thing I knew it landed in Cork," she says.

By all standards, the Irish are a loquacious people, given to emotional debate. In her community, at her post office and in church groups, abortion has given way to genetically altered food as the hot topic. She's heard that nobody is talking in the United States, and she wants to know if that is true. I tell her that in America, people seem to have more faith in their government to protect them.

"This is about our food and what people eat. Americans know that, don't they?" she asks. "Maybe America has just gotten too big for people to talk to each other."

CONTINENTAL CONCERN

By early 1998, Europeans had begun to suffer buyer's remorse over a ruling their continental governing body had made nearly two years earlier. In April of 1996, people had paid little attention when the fifteen-country European Union approved the import of Monsanto's genetically engineered Roundup Ready soybeans.

European newspapers all but ignored the event. So did news media in the United States. All but my newspaper, the *St. Louis Post-Dispatch,* which reported the news from a press release sent out by the American Soybean Association, which is based in St. Louis. From every indication, another routine success by the hometown company.

Europe grows few soybeans, so it was understandable that European papers weren't tuned in. As recently as 1997, biotechnology sprouted as quietly as corn grows. In the United States that spring, the first year farmers could legally plant engineered crops, about seven million acres were sown. Outside of the farm journals, few American media organizations noticed. Soybeans from many fields are mingled at grain elevators for wholesaling, so the shipments reaching Europe in November stood a good chance of containing beans with the newly altered genes.

In February 1997, the European Commission had given another authorization. Now Europeans would not only be importing modified food but also growing it. The crop was engineered corn, and this time the company was Swiss-based Novartis, not Monsanto. Immediately, Austria and Luxembourg banned the Novartis corn from their countries. Norway followed a few months later.

Meanwhile, France was preparing to offer its fields for the first plantings of the Novartis corn. The French parliament was organizing a focus group, and a pollster had been hired to select the citizens and bring them to Paris in June to advise the government. I had made plans to be there because France—as Europe's breadbasket—would be a battleground if a food war broke out in Europe.

In Britain, food became a subject of worry as the tally of deaths grew from B.S.E.—bovine spongiform encephalopathy—the brain malady in cattle that was believed to have killed more than two dozen people. Mad Cow Disease, as Europeans called it, had already cost the country billions of dollars. People were furious over the failure of politicians and regulators to protect them. Not until mid-1996 had the government confessed the whole truth.

And now, critical biotech decisions loomed. In less than three weeks, in March of 1998, the European Commission would vote on whether to approve four more modified crops. If they allowed all these seeds to sprout, genetically altered food would take root on European soil once and for all. For Monsanto and its biotech rivals, Europe is the world's biggest market—and more. Europe's imprimatur could speed acceptance of the technology in former European colonies around the world.

In early 1998, when I arrived on the scene, the atmosphere in Europe was charged, but no opposition had coalesced against genetically modified organisms, or GMOs as Europe had begun to call them. Greenpeace was working to change that. In the United States, Greenpeace has a reputation as a ruckus-loving bunch that had achieved organizing successes on the toxics front but had since lost firepower. But in Europe, Greenpeace had swelled into an environmental heavyweight with the muscle of two million members.

Greenpeace began organizing against GMOs in Europe in mid-1996, but the campaign was not automatic. Internal debate had been spirited about whether to take up the cause. Some members preferred the boat-chasing, banner-hanging tactics that got results because citizens assumed something had to be wrong to provoke such tactics. Fighting GMOs was different than saving whales or banning chlorofluorocarbons in refrigerators. There was no assumption of success, as in the Greenpeace waste-trade campaign that led to the Basel Convention regulating shipments of toxics from industrialized countries to the developing world. In GMOs, Greenpeace strategists saw a monumental task of grass-roots organizing, an effort that could take years and still fail. That was a battle they weren't sure they were prepared to make.

"This was not something that lent itself to the kind of heroic actions we were used to, in which you expose something and you stop it. It wasn't like going to the Arctic or entering Brent's Spar," said Benny Haerlin, a senior Greenpeace campaigner and former member of the German parliament, referring to the distance Greenpeace had gone to stop Shell Oil from drilling in the North Atlantic

"This was a campaign that needed to be thought out on the market level. We knew we had to get as many consumers involved as possible, and that's what we tried to do."

Not even Haerlin, who had kept his eye on genetic engineering, knew how swiftly European foodstuffs would become altered once stevedores unloaded the first engineered soybeans in late 1996. He saw the light when a representative from the Tengelmann supermarket chain in Germany told him that over half of the prepackaged foods on their shelves soon would contain soybeans that might be modified. What would Greenpeace think of that? Haerlin was asked.

Unwittingly, the food industry had spurred Greenpeace into action. Greenpeace campaigners showed up to meet the ships laden with the cargoes of newly modified soybeans. By 1997, Greenpeace had organized 250,000 consumers, many of them writing letters to grocery chains.

In Ireland, due in large measure to Greenpeace, Irish readers would open their *Irish Times* in 1997 to a story headlined: "Frankenstein Fodder or Food for the Gods?"

RESISTANCE, ROMANCE

In Dublin, Quentin Gargan also worried about genetically engineered soybeans arriving from America. He operated his own company, Wholefoods Wholesale, which supplied health-food stores around the country with such staples as soy milk and tofu. To people who have sworn off meat, soybeans mean protein. At Gargan's urging, the Irish Health Stores Association agreed to seek a ban on modified beans.

"From the point of a view of attracting the attention of people who are really into food and health, soybeans was not the way to go. If Monsanto had brought in something else, it might have been different," Gargan said.

Clare Watson also fretted about newly arriving GMOs. In January of 1997, a friend at the Cork Environmental Alliance phoned to tell her that Monsanto had applied for permission to test genetically engineered sugar beets in Irish fields. Watson, who had worked for Greenpeace and for other environmental groups, invited Dublin activists to her apartment in March of that year to talk over what could be done. Among those who showed up was Gargan, whom she had met before at a gathering to discuss French nuclear testing. Out of that March meeting sprang Ireland's anti-GMO organization, Genetic Concern. So, by the way, did a love affair between Gargan and Watson.

Environmentalists in Ireland say their campaigns are hard to sell because their land seems unspoiled. But by the time I arrived in Ireland, Genetic Concern was a spunky yearling. At a Sustainable Earth Fair at Maynooth University near Dublin, people wore black T-shirts sporting the Genetic Concern logo. Gargan and Watson shared equal billing with John Seymour, Ireland's guru of self-sufficiency, and Richard Douthwaite, the radical economist.

Gargan sat on a panel with Patrick O'Reilly, a Maynooth University graduate and Monsanto's business manager for Ireland. Gargan recalled the old slogan, "DDT is good for me," as a way of saying that the government of the United States can be wrong. O'Reilly's jeering reception marked a rude homecoming for a successful alum.

"How are we going to feed the world?" O'Reilly began, deploying an argument that the biotech industry expected to resonate in Europe.

But on the floor of the drafty Maynooth hall, many in the audience were unpersuaded. Indeed, I heard people talking openly about destroying Monsanto test plots.

Threats didn't sit well with O'Reilly. "Sabotage is a criminal act," he said.

SABOTAGE

"It's 2AM, a fog has descended and I'm arching my back after an hour of being bent over working. I take in what I see around me. Set against the yellow lights from a nearby industrial plant, a figure lobs another destroyed vegetable through the night. Nearer to me, another darkened figure slashes at the earth with the glee of a pixie . . .

In the twenty-acre field I am standing in, a whole acre of sugar beet is being systematically destroyed. As the sugar beet is hacked at, ripped out, slashed and beheaded, this doesn't at first sight resemble what I thought an 'action' would look like. In fact, it looked more like the world gone wrong, the people fighting for the planet were now attacking it with vigor. The cobblers' fairy tale of pixies visiting during the night to finish off the work comes to mind. But we were not completing someone else's work, we were completely destroying it . . .

Despite protests and calls for a debate on the issue, an American multinational company has pursued its global strategy concerning the development and introduction of new foodstuffs—genetically modified organisms . . . They have refused to inform the public, have refused to acknowledge any possible long-term effects from the introduction of GMOs. They have refused to play ball and tonight a local group of ordinary people are going to stop them . . .

At the end of the action, there was no sudden sense of great achievement and no real celebrations. Instead, there was something simpler: a feeling that we came, we dug and we made Ireland GE-free again."

So wrote an activist from the underground Gaelic Earth Liberation Front, after a midnight visit in late 1997 to Monsanto's test plot at a government-owned research center in County Carlow. It was the boldest attack yet on a field of gene-altered crops, and it was especially destructive because of its timing—the day before the sugar beets were to be harvested. Monsanto was on the verge of reaping the fruits of a long season that had begun with the public battle to win Irish approval for the field trials. The attack—but not the identity of the saboteurs—was first reported by militant environmentalists on the Internet, the new tool of revolutionists everywhere.

"That went totally against Irish culture," Monsanto's Patrick O'Reilly told me in Dublin.

ROOT VEGETABLES

If you're going to appeal to the Irish, it makes sense to talk potatoes. At Maynooth University, Patrick O'Reilly talked Monsanto's NewLeaf potato, engineered to ward off disease while allowing the farmer to cut back on pesticides. It was already sprouting across the United States. "If this is not a clear example of sustainability, I'd like to hear a better one," he said.

When the Irish have too few potatoes, there's trouble. A partial failure of the crop in 1727 led to riots in Cork. But nothing compared to the catastrophe of sixty-two years later. The summer of 1845 seemed promising for Ireland's potato crop. Early varieties escaped the blight spreading in Europe. But by October, just-picked potatoes rotted. The next year, the blight devastated areas that had escaped the year before. A country parson named M.J. Berkeley discovered the cause of the blight almost immediately—a fungus spread by spores riding on a humid easterly wind from France to England to Ireland.

The failure had its roots in crop genetics. At the time, Irish farmers planted just two potato cultivars. Both proved to be susceptible to the fungus *Phytophthora infestans,* also known as late blight, which spread because of genetic vulnerability. Such a fungus probably couldn't take hold today because of modern fungicides. And if it struck, farmers could plant other varieties of potatoes traded in the modern world. Nonetheless, the Great Famine colors the way the modern Irish think of food.

Monsanto's O'Reilly said that his company would have preferred to plant modified potatoes to modified sugar beets for Ireland, but potatoes weren't practical. If they'd started with potatoes, they'd have had plenty to choose from. On March 2, 1995, the United States Department of Agriculture approved seven lines of Monsanto potato engineered with the Bt gene for resistance to the Colorado potato beetle. More approvals have been awarded since, and Monsanto's NewLeaf potato was at the time in production on tens of thousands of acres in the United States.

Monsanto has experimented even more broadly by reconstituting potatoes with several new genes, often at the same time, I learned from applications submitted to the government. In these "stacked-gene" experiments, as many as eight traits were engineered into a single plant: genes for resistance to pests and diseases, for tolerance of direct application of herbicides, for more solid content, and for less bruising.

In Ireland, the company began with just one crop with just one engineered trait: herbicide-tolerant sugar beets, which hold vast commercial potential.

In Babylonian times, beets were raised for their leaves. Later, they were developed for their roots. And then, in the eighteenth century, for their sugar content. Now, roughly one-third of the world gets its sugar from sweet beets the size of softballs. (Most of the rest get it from sugarcane.) In Ireland, just eighty thousand acres are sown in beets. But the two hundred thousand tons of sugar from those acres make sugar Ireland's most enduring cash crop. What Monsanto has done to the genes of the sugar beet is just what it has done to modify soybeans, corn, and potatoes in the United States: inserted genes to create herbicide tolerance. The plant's physiology has been altered so that growers can liberally apply Monsanto's Roundup herbicide—or a competitor's similar glyphosate formula—over the top of plants, killing weeds without fear of harming the plant.

Herbicide tolerance is the most popular trait-splice in agriculture. But it has won biotechnology companies few friends among Europeans, who see no benefit to themselves. In Ireland, a sprawling wetland where crops perennially fall victim to weeds and diseases, genetically engineered herbicide tolerance can appeal to a farmer. As it does to farmer Richard Fitzgerald.

A FARMER STEPS FORWARD

With sabotage in the air, Irish farmers are wary this season of offering their land for field tests. Richard Fitzgerald is an exception. One of five Irish growers who have agreed to work with Monsanto, he invites me to visit

Sugar-beet grower Richard Fitzgerald dared to turn over
his land to genetic-engineering field trials. He joked that
he might need assistance from fellow farmers "in case
little men grow out of the ground where they plant those
seeds."

his farm near Loughane in County Cork. There is nothing much to see
but fields of rich, furrowed earth and mounds of sugar beets from Fitzgerald's last crop. But Richard, an affable man with a wicked sense of
humor, makes my trip worthwhile.

Richard is in his late forties, and his florid face beneath his blue Nike
cap bespeaks his love of Guinness. The prospect that his beloved Guinness
will be brewed with Monsanto's gene-altered barley does not trouble him.
Richard, a pioneer in beet-growing, has a reputation for innovation. As
an officer of the World Association of Sugar Beets and Cane Farmers, he
has traveled as far as Thailand in search of new tricks. People respect his
opinions and look to him for guidance. So his fellow beet-growers did

not object when he stood at a meeting of the Irish Farmers' Association's Sugar Beet and Vegetable Section to declare his intention to test the new crops.

Richard is the sort of well-heeled, establishment farmer the biotechnology companies have cultivated in the United States—and hope to win over in Europe. From his BMW sedan, he scans his two hundred sloping acres near Ireland's seacost. In late winter, they look like an empty chessboard. Richard relishes the order of his terrain and points with special pride to his clean, deep brown fields. Herbicide spraying keeps them tidy, while nearby fields are fuzzy with weeds and volunteer cereal grains. Richard doesn't abide weeds. Monsanto will plant about five thousand engineered beet plants on a quarter acre of his land—if Genetic Concern's court fight or anti-GMO avengers don't get in the way.

"If it turns out to be good, it would mean that you can control your weeds," he says of the new gene-altered sugar beets. "Let's be clear; there is no proof that it will be good for us. And the seed might be expensive. But we have to try it."

Richard claims not to fear a midnight raid like the one at County Carnow that prompted half the farmers to back out of Monsanto's trials. But he doesn't look fondly, he says, on the prospect of people traipsing about his land. Or their motives. "When they talk about organic, who's going to feed the growing population?" he asks. "It's not going to be a few environmentalists with a couple acres of potatoes."

But Richard must attend to public relations. That's why, after a sandwich with none of the Guinness he has forsworn for Lent, he must hurry to a meeting of neighboring farmers, where he will be rounding up more support for the field test on his land. When I ask why he needs reinforcement, he winks at me: "In case little men grow out of the ground where they plant those seeds."

BALLYMALOE COOKERY SCHOOL

A neighbor whose support Fitzgerald never will secure is Darina Allen. With her best-selling cookbooks and her television shows, Darina Allen

At her Ballymaloe Cookery School in County Cork, chef and best-selling author Darina Allen advises students to steer clear of genetically modified ingredients.

is Ireland's Julia Child, with a dash of Martha Stewart thrown in. But down at her famous Ballymaloe Cookery School at Shanagarry, adjacent to Fitzgerald's acreage, Allen's normally cheery outlook is dampened these days by something other than fallen soufflés: genetically modified food.

With few exceptions, you don't find prominent American politicians and icons speaking out against genetic engineering. Not so in the United Kingdom where, in early 1998, Prince Charles is rumored to be on the verge of condemning the technology and Sir Paul McCartney, the ex-Beatle, is thinking likewise. Darina Allen may not come from royalty or rock 'n' roll, but she is a home-grown star in Ireland. It would be hard to find someone in Ireland who hasn't seen, heard, or read Darina Allen or journeyed to Shanagarry to eat well at the family's Ballymaloe House restaurant.

She has sold more than eight hundred thousand copies of her Simply Delicious cookbooks since the first best-seller in 1989. Her television cooking shows are syndicated as far away as India, and her school has a waiting list of pupils eager to enroll in twelve-week culinary courses. She sits on Ireland's governmental Food Safety Authority. Whatever stage she occupies, Darina Allen lectures the Irish to buy fresh, natural foods to achieve both the best taste and good health.

That's her policy, she tells me as we tour her organic farm. "I am forever telling my students, 'this is the person who raises your chickens. This is the person who caught your salmon.' I think that in the kitchen, they must see the faces of the people raising the food." She is attired in her white double-breasted chef's smock with a white apron and black leggings, and her hands seem ever affixed to her hips, as if she's always about to expound. She's wearing prominent eyeglasses that have become a trademark; in a comedy skit on Irish television an actor identified as "Darina Allen Genetically Engineered" appeared on the screen wearing huge glasses with lenses the size of sugar beets.

The Ballymaloe Cookery School that Darina and Tim Allen operate sits on one hundred acres near a fog-swept Irish seashore. Besides the organic farm that produces everything from herbs to eggs to free-range chickens and meat from Kerry cows, it has an acre of greenhouses. Box-woods and gardens ring the estate. Darina is carrying forward a worldview sown in the Irish consciousness by her mother-in-law, Myrtle Allen, who had been Ireland's most revered chef. More than fifty years ago, Myrtle and her husband, Ivan, who were Quakers, bought Ballymaloe House, a Georgian mansion, opening first a restaurant and then a country hotel.

In the vestibule of Darina Allen's school, I become disoriented. Brightening the solid surroundings are walls the color of limes, oranges, and orchids, along with papier maché animal statues, which I recognize as folk art from the Sierra Madre mountains near the central Mexican city of Oaxaca. Yes, Darina Allen tells me before inviting me to lunch, she vacations there.

Eating a dish prepared by her students, salmon encrusted in pastry, in a dining room overlooking a courtyard, I conclude that Darina Allen must indeed have been cloned, as the comedy skit suggested. She's leaning over me. Then she's in the kitchen sampling and lip-smacking. Now she's at the table eating. Then I see her at still another table in front of a different plate.

She has a disarming bluntness, asserting simply and firmly that we ought not be "fiddling around with genes." It is the next generation of thinking beyond her mother-in-law Myrtle Allen's crusade against modern food-production techniques, what Myrtle called "this constant, pressure, pressure, pressure for cheaper, poor-quality food."

Darina Allen minces no words in decrying "global food rot" from genetically modified ingredients. In her view, the rotting of soybeans is nearly complete, and the next staple to fall victim will be corn.

"What concerns me," Allen says, "is that so many genetically engineered foods are in the foods we eat. It worries me enormously. I can see how incredibly exciting these developments are for scientists. But this is a serious danger, a sinister monopoly. It is a basic human right to be told what's in your food.

"I'm not a romantic; I'm a country girl. In my simplistic way of looking at things, you can't change nature and continue down this path of intensive farming without paying a price."

For the biotech industry, Darina Allen is a formidable foe. But a greater threat is what she represents—Europe's multibillion-dollar organic-food industry and the growing sense that genetically modified food has nothing to offer people who equate health with what they eat.

QUENTIN AND CLARE

While underground activists plot and Darina Allen inveighs against unseen threats, Clare Watson and Quentin Gargan labor to sway public opinion against the sugar-beet tests. Their Genetic Concern is one of a kind in Ireland—and one of few groups anywhere in the world—organized solely against genetic engineering.

Gargan, who was born in 1956, had grown up in the food business; his family's J.M. Gargan and Company, started by his grandfather, operated a sweet factory, a sweetshop, and wholesale business. Its wares were antithetical to the foodstuffs that Quentin began to sell when, at twenty-five, he set up his own wholesale organic business. "I never was very happy that my day's work consisted of putting a ton of sugar into the food chain," he said.

Watson, who has cascading brunette curls, was born in Kenya in 1962 while her father was managing a tea estate. She is trained to work with disturbed children but gravitated to full-time pro-environment pursuits. And full-time is what Genetic Concern turned out to be—sixty to seventy hours a week for both her and Gargan, who was still operating his food business. Their relationship flowered in the months after they established Genetic Concern. By May of 1997, two months after Clare convened a handful of environmental advocates, they were together.

They are a handsome, earnest couple, and they give the fledgling Irish opposition to genetic engineering an unthreatening public face. They're plugged into an expanding anti-GMO network globally, deploying the Internet to strategize and share tidbits of information gleaned by activists around the world. And they are skilled at working with reporters, judging by the attention they received in Irish newspapers.

From their headquarters on Dame Street in Dublin, Genetic Concern distributes loads of brochures, some factual, some less so and some alarmist. A flyer entitled "Nature Knows Best" has the message: "Your food is being tampered with. Were you consulted?" It also has misleading passages such as "Crossing Cows with Cabbage."

By most accounts, Gargan and Watson are effective. While I am in Dublin, the Irish Food & Drink Industry announces at a news conference that it will call for voluntary labeling of products and ingredients from modified crops. It is a preemptive strike against public opinion that is beginning to run against genetic engineering, thanks to publicity fueled by Genetic Concern, Darina Allen, and the Green Party in Ireland.

In my mind, there is no better measure of the activists' success than an uncommon alliance that sprouted in opposition to GMOs: both Sinn Fein President Gerry Adams and Ulster Unionist Party leader David Trimble have denounced the testing of gene-altered plants. In genetically modified food, Ireland's warring factions found a common enemy.

But when I meet with Clare and Quentin for dinner in Dublin, I wonder how long they will stick with their campaign. I can see they're in love, but their organization seems frayed. They have forged a membership of over five hundred people who have contributed money to keep them going. They are winning sympathy, but it is harder to attract committed people at a time when Ireland's economy is booming. There is the matter, too, of Gargan's conflict-of-interest: His multimillion-dollar organic-food business stands to profit if Irish consumers reject the engineered, processed foods from intensive agriculture.

The main threat to their organization—and perhaps to them—is their court campaign to derail Monsanto's experimental plantings. In Ireland's High Court, Watson has challenged the award of a field-test permit first given to Monsanto in 1997 by the country's Environmental Protection Agency. She won a temporary injunction, which later was lifted to permit Monsanto's ill-fated planting in County Carlow. Now, Watson is hoping that a High Court Judicial Review will be completed in time to forestall the spring sugar-beet seedings on Richard Fitzgerald's County Cork land and at four other Irish locales. At dinner, I learn that this court action was brought not by Genetic Concern as a group but by Clare Watson as an individual. Under Irish law, the loser is responsible for the cost of proceedings. In this high-profile case, those costs could be heavy.

Watson tells me that she knows the stakes. "The barristers sat me down and told me what might happen. They said you'd better go outside and have a think before you go through with it. I did, and when I came back, I said that this is something I have to do."

As we talk, I have the sense that Clare and Quentin have not fully considered the consequences of their activism.

THE SAGA OF BERTIE AND BERGER

I have watched Monsanto operate for over twenty years, and I have learned not to underestimate the company's capacity at lobbying. Wherever it operates, Monsanto doesn't hesitate to bring on public affairs muscle and to throw money at lawyers, experts, and the political elite. I assume that shareholders would expect no less.

In the late 1980s and early 1990s, the company's skills and contacts in the United States cleared the way for the most successful introduction of a new agriculture technology in history. The company's tenacity at manipulating bureaucracy—and its blockbuster Roundup herbicide—were in large measure why it was so successful. In Europe, Monsanto was still learning when to use muscle and when to finesse. In Ireland, some of the company's maneuverings were on display and some, I learned, were not.

As the Irish debate was brewing, the company flew a group of Irish journalists to the United States for a tour of its St. Louis labs. The tour included a stop in Washington, where the journalists received a bonus: They were escorted into the White House for a visit to the Oval Office. "Our little heads peeked around that historic room," recalled Vivion Kilferther, a reporter for the *Examiner* in Cork. Few White House visitors see the inner sanctum of the most powerful office in the world. But if you're Monsanto, and you've hired away the president's director of intergovernmental affairs, Marcia Hale, you may be accorded special treatment.

In the Irish High Court case, Monsanto also labored diligently. Resistance to its field trials, an affidavit filed by a company lawyer warned, could mean problems for Irish farmers in getting not only modified seeds but also the traditional seeds that the sugar-beet industry needs to carry on. "This would have serious implications for the Irish sugar-beet industry," the affidavit read.

While waiting for the High Court to rule, Monsanto fixed its tactics on bigger stakes—the pending decision by the European Commission on its corn and several other products. The Commission was due to make that decision in March of 1998, and the support of Ireland in a vote weighted by population was critical to the outcome. On St. Patrick's Day, during a visit to Washington by Irish Prime Minister Bertie Ahern, Monsanto set a plan in motion.

When Ahern arrived in Washington, he was greeted by National Security Council director Samuel "Sandy" Berger. At lunch in the Capitol, they talked not just about the Irish peace talks or any of the world's flare-ups that typically sent the rumpled Berger scrambling in front of CNN's cameras. The issue was Ireland's pivotal vote in the European Commission on genetically modified crops. Later, a National Security Council official offered me an explanation for the meeting. "In this post–Cold War era, America's national interests have changed, and crises aren't always military crises," he said.

Berger's wasn't the only voice Bertie Ahern heard in Washington. Senator Christopher S. Bond, a Republican from Missouri, and several members of Congress collared Ireland's prime minister, as did others in the Clinton administration. Toby Moffett, then Monsanto's Washington-based international business head and a former Democratic congressman from Connecticut, marveled at the smothering of Ireland's prime minister.

"Everywhere he went, before people said 'happy St. Patrick's Day,' they asked him, 'What about that corn vote?' " Moffett said. "I'm fifty-four years old, and I've been in a lot of coalitions in my life, but this is one of the most breathtaking I've seen."

The next day, the European governing body approved Monsanto's plantings of gene-crossed corn and three modified crops of rivals. The votes are secret, but Irish officials later said that they'd voted for a corn manufactured by AgrEvo and abstained on the other three products. It was the first time that Ireland had supported release of a genetically engineered corn product. The Commission's decision was good news for Monsanto and for American farmers switching to modified crops. And it turned up the heat on France, which would have the last word on Monsanto's corn.

When I reported the pressure on Ahern later in 1998, my story had an impact that I hadn't foreseen when I wrote it. The impression it left that Ahern "got rolled," as they say in the vernacular of Washington, gave ammunition to GMO and Ahern administration critics alike.

News stories in Ireland carried criticisms of the government for bowing to pressure from the United States. The issue stewed for months, finally boiling over on the floor of the Dail in October 1999, when Environment Minister Noel Dempsey was grilled by Labor Party member Eamonn Gilmore on the details of the meetings between Bertie Ahern and Sandy Berger and what the Ahern government did afterward.

Gilmore: "Does the Minister think it was appropriate to instruct his officials to vote in favor of the release of genetically modified organisms on foot of lobbying from the U.S. government at a time when he was supposedly engaged in consultation with a range of interests in this country to formulate Irish government policy?"

Dempsey: "It was as appropriate to instruct an official to vote for it as it was to instruct an official to vote against it or to abstain."

Gilmore: "Was the Minister not pre-empting the entire consultative process over genetically modified foods by voting in line with the lobbying which had taken place the day before between Mr. Berger of the United States and the Taoiseach?"

Dempsey: "The deputy's argument does not hold water."

Gilmore: "The Minister was leaned on to vote in favor of it. He made a joke of it."

Genetic Concern, the advocacy group which was fighting now to stay afloat, responded with a news release headlined: "Dempsey owns up to U.S. lobbying on GM crops." I noted that rather than Quentin or Clare, a woman named Sadhbh O'Neill was speaking for Genetic Concern. She said:

"This is the first time the Minister has candidly admitted that not only did the U.S. National Security Advisor lobby the Taoiseach last year in relation to GM crops, but that the Taoiseach instructed Environment Minister Noel Dempsey to vote in favor of a GM crop at a crucial E.U. meeting on the eighteenth of March, 1998. . . . We were promised a moratorium, and we got a government that buckles to U.S. pressure. We were promised a consultation process, and we got a sham. . . . U.S. multinationals have more influence than the Irish electorate."

The debates testified to the political currency of the GMO issue in Europe. It was nothing like what you would have heard in the U.S. Congress, where food biotechnology still enjoyed bipartisan support. I found the hullabaloo amusing because it is my belief that the Monsantos of the world exercise behind-the-scenes power every day. Monsanto wasn't amused. A company official told me that my revelations in print had jeopardized Monsanto's ability to win Irish backing, making it harder to win approval in the European Commission.

But by then, other forces gathering across Europe had dwarfed the biotech problem in Ireland.

EPILOGUE

By the summer of 2000, Monsanto's drive to cultivate modified sugar beets in Ireland was headed in reverse. Nearly three years after planting ten field trials as a prelude to commercialization, Monsanto's bioengineered sugar beets were sprouting in just two Irish test plots. The government was nowhere near legalizing gene-altered seeds for use by farmers.

When I telephoned farmer Richard Fitzgerald, I found him recovering from a serious farm accident. Stepping into a piece of machinery hooked to the back of his tractor, he fractured his right leg in two places and spent nine weeks in the hospital.

His experimental sugar beets fared worse. In August 1999, saboteurs struck his test plot of Roundup Ready sugar beets. They hacked down about half of the plants with machetes. After Monsanto officials surveyed the damage, they destroyed the rest of the plants under the supervision of Ireland's EPA. No arrests had been made.

Nor were any members of the Gaelic Earth Liberation Front, who carried out the County Carlow destruction in 1997, apprehended. Of the four incidents of sabotage in Ireland since 1997, attackers had been prosecuted in only one. Six of the "Arthurstown Seven" were found guilty of criminal damage but received only small fines. A headline in the *Irish Times* about the convictions did not please Monsanto. It read: "Saboteurs with 'honestly held beliefs' cheered."

Richard said that the tests with herbicide-tolerant sugar beets showed that they would grow with fewer chemical sprayings. From that standpoint, they were successful, he said.

When I asked Richard about the attitudes toward GMO these days, he replied: "There's a lot of anger. People are just standing back a bit."

The accident hadn't damaged Richard's piercing wit. "I guess there are people around who think that if you eat the sugar from these little plants, you'll grow breasts," he joked.

While convalescing, he enjoyed a bouquet of sympathy flowers from his neighbor, Darina Allen, who continued to be Ireland's most famous anti-GMO celebrity.

I telephoned Mairie Cregan who, despite a household of ten children, went back to college. Her husband, David, got a job with a computer company in Cork. They had kept the farm going, and they were milking even more cows since son Rory, who had turned eighteen, worked with them on the farm.

Mairie said that GMOs had been in the news every day since recent revelations that modified rapeseed slipped into Europe and was unknowingly planted by farmers. I asked her if people were still debating the technology, as they were two years before, and she answered yes.

"Some people are a bit pissed off that we don't have labels yet. Others are getting frightened when they hear how genes can be transferred between species. I think that some people are going to choose to buy organic. But for other people, just getting through the day is hard enough for them," she said.

I asked Mairie if she believes that consumers can continue to block the wholesale plantings of modified foods. "I think it's going to happen," she said of the arrival in Ireland of modified crops. "But I don't think people are going to make it easy for them."

Meanwhile, Genetic Concern no longer existed. In late 1999, Quentin Gargan and Clare Watson left Dublin, the organization they founded, and political activism. They bought a thirty-four-acre farm in County Cork, where they were growing vegetables, building a greenhouse, and raising sheep and bees.

Watson lost the High Court decision, which allowed Monsanto to proceed with the sugar-beet trials and left her owing about $650,000 in court costs. The case had weighed heavily on her. But speaking over the sounds of a bawling, bottle-fed lamb, she said she was feeling better.

"I'm down to much more practical things now. It's a problem for me if we don't win this, and it will depend on how the EPA and Monsanto deal with it. Obviously, it affects my long-term financial security. I went through a phase of saying, 'what have I landed myself in?' But I've come out of it. We have sea views here and sunny weather, mostly," she said.

Genetic Concern hung on for a few months after its founders left, then disbanded in April 2000. Gargan said that he and Watson did what they set out to do: develop awareness of what was happening in food and farming outside of the public view. He takes credit—fairly, I think—for triggering a debate across Ireland.

In the end, it came down to personal considerations for Gargan and Watson who, it can be said, have Monsanto to thank for finding each other. "I just decided there's no point in running around campaigning and trying to look out for things for everyone else. We always wanted to live in the country," he said.

13

IN FRANCE, DEMOCRACY EUROPEAN STYLE

When I'm in France, I know damn well where the food is coming from because I bought it from a farmer this morning.
—John Richardson, British diplomat

How can you govern a country with three hundred varieties of cheese?
—Charles de Gaulle

DAY ONE: "ANTI-AMERICA . . . ANTI-MONSANTO"

AN HOUR INTO Europe's newest exercise in democracy, French authorities wonder if their grand citizens' conference on genetic engineering is a mistake. Claire Falhon, twenty-nine, who is partial to plunging necklines and who described herself to me as "passionate in my body and mind," is shaking a finger at a Monsanto scientist. "You're very interested in selling your products, we know that. But what about getting advice from our doctors first?"

Falhon is one of fourteen French citizens chosen randomly from across the country to guide government policy on a conflict embroiling Europe. This is supposed to be a "consensus conference," where people listen, politely debate, and ultimately resolve their differences. But in Paris, the proceedings organized by the French National Assembly sizzled from the start. A French scientist billed as an impartial expert has showed up for his panel in a black-leather Guns 'n' Roses vest and an attitude to match. And this Falhon woman looks as though she's about to go over the edge.

In their blue-leather armchairs, members of parliament trade nervous glances.

The French government had hired a pollster to select fourteen people from the nation's population of sixty million. For three weekends leading up to today, these special citizens were delivered to Paris for science lessons so they wouldn't be confounded when the conference began. They were cloistered, their identities concealed from the French news media and from advocates on either side eager to sway them. Now, in front of fifty reporters and cameras, the pot is boiling over.

One of the citizens has just grilled the experts from government and industry on what happens when GMOs are released in the environment. Does altered DNA break down? Does it remain intact? Reasonable questions, as I see them.

"We are all GMOs that degenerate," a scientist replies, referring to the human body.

Claire Falhon considers that answer patronizing. "That's nature; we're not talking about nature," she interrupts. "Don't mix up things between nature and what we modify."

Falhon rejects the claim that French farmers need genetic farming to survive. "It's not just the right of farmers to have a choice; it's our choice, too," she says.

Other citizens jump in. "What will happen in ten or twenty years is the fear of the people," says Michel Martinet, a dentist from wine country. "The problem is that you're putting out genes that can spread all over the world."

Francine Maeght, a bank worker, concludes that genetic engineering threatens the environment. "If a GMO plot reproduces itself, is it possible that nature escapes our control because we have acted as sorcerers' apprentices?" she asks. "In my opinion, this is a very important risk to take. I want to protect the diversity of the French landscape."

The bevy of industry experts on hand expected pointed questions. They're aware that guerilla resistance is breaking out across Europe. But they haven't anticipated the early heat, and nor did they expect attacks from scientists in the French government. The French Environment Ministry's J.L. Pujol, a panelist, predicts lasting damage from genetically modified seeds. "Contamination of the soil is irreversible," he says. He wants to put on the brakes. "There are too many risks, considering the benefits," he says. "We have to slow down because we cannot reverse the procedures."

Next, Pierre-Henri Gouyon, a University of Paris professor, advises a moratorium on planting while French scientists conduct their own studies. Gouyon, a rotund fellow whose middle swells his Guns 'n' Roses vest, identifies himself as one of only six people in France who study risks. Many more work in the service of production, he adds.

Then Guy Riba, who works at a French government research agency, complains that modified food is being forced on his nation. "The position of the United States is unacceptable to us," he says, criticizing the U.S. government's trusting habit of accepting the commercial seed-makers' test results when setting regulations. "We have a very good agriculture system. Public research needs to be developed. Otherwise, the system of control would be left to international companies."

The French don't, by nature, look favorably on the United States. After World War II, it was fashionable in Left Bank cafés to minimize the United States role in the emancipation of France and to view the Marshall Plan as a blueprint for control. Multinational companies of American derivation shared the rap. When Coca-Cola came to France in the 1950s, protesters took to the streets chanting *coca-colonization*.

At the end of the first day of discussion, Monsanto's Stephane Pasteau, a veterinarian, looks crestfallen. Even the government scientists are talking about a moratorium on genetic science. Retreating from this day's loss, Pasteau shakes his head ruefully as I take him aside. "I do not have a good feeling," he says. "It is so anti-America. So anti-Monsanto." He is praying for better treatment on Day Two.

TROUBLE IN THE COUNTRYSIDE

The French conference is more than a simple exercise in democracy. What they decide will help shape French policy that could in turn determine the future of genetically modified food in all of Europe. And around the world, nations are waiting to see how Europe reacts.

Until recently, the arrival of genetically modified food in Europe looked like a *fait accompli*. Three months earlier, in March 1998, the European Commission approved plantings and import of corn and other modified crops. The Commission's yea should have opened Europe's door, a giant and seemingly irreversible step for biotechnology worldwide. But under Commission rules, another hurdle had to be cleared. The country that had sponsored the corn applications, the *rapporteur* country, had to sign off. That country was France.

Now, the very country that pronounced the corn safe was balking. Pressure from neither biotech companies nor the United States government availed. Single-handedly, the stubborn French were blocking shipment of $200 million worth of U.S. corn to Spain and Portugal.

France is no nation of Luddites. The country gets much of its electricity from nuclear power plants. Nor is France known for coddling environmental crusaders. Greenpeace discovered as much in 1985, when an explosion sank its *Rainbow Warrior,* then docked in Auckland, New Zealand. A photographer on board died. The French government, angry at Greenpeace's interference with nuclear weapons testing at Mururoa Atoll, had ordered bombs planted on the Greenpeace vessel after other attempts at sabotage failed. Two French agents were sentenced to ten-year prison terms.

Now, environmental advocates around the world had their eyes on France, as the socialist government of Lionel Jospin debated the call. France had seemed all for biotech. In November, that nation had agreed to host the first planting in the fifteen-country European alliance of modified corn, an insect-resistant hybrid sold by Novartis.

That trail-blazing planting didn't go according to plan. Novartis had predicted that seventy thousand acres of Bt corn would be sown. But when planting season ended, just four thousand acres had gone in. The company's rosy hopes had been all but destroyed on a January afternoon when farmers stormed a Novartis storehouse near Toulouse in southern France and destroyed millions of seeds. When I asked Novartis's Philippe Gay about the attack, he shook his head and exhaled. "It was pretty nasty," he said.

On that day, January 8, 120 farmers had assembled in the center of the town of Vallonjue. They traveled in a caravan of trucks and cars to the Novartis storage facility in the town of Nerac. Leading the protest was René Riesel, a sheep and grain farmer who was also the national secretary of Confederation Paysanne, France's second biggest farmer's union.

When they got out of their trucks, the farmers barged into the Novartis building. "Where do you store the damn genetically modified seeds?" Riesel demanded of a worker.

Six months later, Riesel replays the attack for me. A wiry, bearded man whose fierce grip belies his diminutive size, Riesel is eating steak tartare and a pile of french fries at the Concorde Café near the French National Assembly. He is wearing sandals and a brown-denim shirt that hangs out over a black T-shirt and blue jeans. As he eats, he smokes nonfilter cigarettes one after another.

"I had a discussion with the boss, and he was very anxious," Riesel remembers. As they spoke, about thirty of the raiders prowled the building. When they found seed bags, "things began to get a little hotter," as he puts it.

The farmers slashed the bags, spilling corn seeds onto the floor. To illustrate what they had done, Riesel interrupts his smoking and eating to pull out his pocketknife. He slashes the air.

When the cutting was done, farmers snatched fire extinguishers from the wall and sprayed the floor with foam. One of his associates told me later that some of the farmers unzipped their trousers and urinated on the mess.

"It was no longer natural anyway, so we wanted to perfect the denaturalization process," Riesel says.

Their trial in February became France's first national confrontation over biotechnology. Environmental advocates and skeptical scientists challenged France's approval of the Novartis planting. Among those testifying for the defense was Vandana Shiva, the physicist from India who travels the globe to resist GMOs. Riesel and two fellow saboteurs were convicted, given suspended sentences and ordered to pay damages of more than eighty thousand dollars. Novartis contended that the destruction merited much more.

The true costs were incalculable. The attack in Europe's primary corn-growing lands was a crunching setback for the biotechnology industry. The sabotage and trial were well reported, and French seed cooperatives that had intended to sell the Novartis seed changed their minds. The French government's thinking was changing, too. In Paris, on the day of the trial, Agriculture Minister Louis Le Pensec acknowledged that the promised national debate on the biotechnology had been "rather limited." Planning began in earnest for the citizens' conference in the French National Assembly.

"We think it was successful because of all the publicity we got," Riesel says. "We wanted to spark a debate among farmers, and we did."

In the months that followed, Riesel went through changes of his own. Instead of tending sheep, he was making speeches across France and distributing photocopies of his anti-GMO manifesto. His aim, he told me, is enlisting support not just from farmers but from broad segments of the French population.

"When you're dealing with questions this important, the whole society must make the decisions," he says.

DAY TWO: WARMING UP

Around the world, the advance of biology is challenging democracy's capacity to make collective decisions. Outside North America, genetically modified farming thrives most where democracy thrives least. Argentina, never a bastion of open government, ranks second in acreage of modified crops behind the United States. China, whose repression is well-known and lamented, ranks fourth behind Canada.

Citizens' conferences are one way European nations have tried to come to grips with this powerful technology. Denmark and Britain had convened earlier conferences, but nothing on the scale of what is now happening in France.

In the United States, I doubt that elected officials, with their consuming desire to remain in control at all times, would risk spawning such an unpredictable gathering. I have seen dozens of advisory panels in Washington, some of them convened in recent years to consider biotechnology.

René Riesel testified at the French National Assembly in
Paris at a consensus conference on GMOs organized by
the government. Earlier, Riesel led an attack on a
storehouse of GMO seeds in southern France.

Almost always, these advisors are careful professionals who follow an
agenda scripted to avoid surprises. They behave themselves, rewarded by
the addition of such-and-such government advisory panel to their résu-
més—and the hope they may be invited back to the nation's capital. Perhaps in cherry blossom time.

When the French reconvene the conference in Paris, Jean-Yves Le
Deaut, a member of the French parliament and president of its science
office, tells the fourteen citizens: "This is a return to democracy, and you
are our grand jury of experts."

On Day Two of the conference, I am prepared for chaos. René Riesel
and Novartis's Philippe Gay are seated near one another on the same
panel, which also includes Greenpeace's Arnaud Apoteker, an avowed
opponent of GMOs. This combustible trio could ignite fireworks among
the French citizens, I fear.

In two decades of political reporting, I've seen what can happen in a focus group. Bring the bunch together a second or third time, and they get weird on you. After their neighbors and friends see them quoted in the newspaper, they turn into performers rather than common citizens. They're experts now, "trained chimpanzees," as a colleague of mine refers to them.

But there's another human impulse that can rescue such gatherings: desire to perform a service. The French citizens received expense francs but no pay for spending their three science weekends away from their families, France's winsome springtime, and the World Cup soccer matches unfolding in their country—with France fielding what turns out to be the world's best squad. Now, on this fourth weekend, they are under searing public heat, lectured and probed on a matter of baffling complexity.

"We owe it to our republic," Georges Schirm, a farmer from the Bas-Rhin region, tells me as the citizens' conference resumes.

When pro-biotech forces learned Schirm was a farmer, they expected to find a sympathizer with a technology that promises to simplify the back-breaking work of weeding and give humans an edge in the endless war against insects. But in Georges Schirm, fifty, they found a sharp and persistent analyst.

"In the end, who decides if Monsanto and Novartis can change the profession of farming? Who decides if they should do this?" Schirm wants to know when the panelists get to ask questions.

He asks, too, why companies need to patent living things. "You don't create anything; you just manipulate the nature of life," he argues. I am so struck by the insightfulness of that question that I fail to write his words in my notebook and must retrieve them later from my recorder.

Novartis's Philippe Gay has an answer. "It is a fundamental right we have to protect our invention as our discovery."

But Gay is also humble. "We haven't communicated enough with the public, which is afraid of our technology even though safety is no problem," he says.

Riesel behaves. He attacks Gay with reason, not his knife. "Just because it has advantages to your board of directors doesn't mean it has advantages to mankind," he says.

Apoteker inveighs against the technology, telling the French citizens that the "Great Satan" in the United States—Monsanto—finances the research that has concluded the products are safe. "They control the genetic growth of the planet. That's too much power for these multinational corporations," he says.

But on Day Two, I see that the French citizens want answers rather than opinions. Marc Planche, an insurance inspector, wants to know how developing nations stand to gain from modified crops.

"GMOs might free us from dependence on the chemical industry," says Didier Marteau, representing the National Federation of Seeds and Agriculture. He tells the citizens that the industry can't afford the moratorium on genetic engineering that many French want.

These Frenchmen and -women are not so aggressive as the day before. As experts describe potential benefits of the new technology—feeding people, raising healthier foods, and curbing environmental damage—I can see the citizens' thinking evolve.

Jean-Yves Le Deaut, the member of parliament running the conference, adjourns before evening to give the citizens a few hours to write their recommendations.

A BATTLE FOR PUBLIC OPINION

Just before the French gathered, the biotechnology industry had won a European victory. In Switzerland on June 7, 1998, voters rejected a ballot proposal that would have severely limited genetic research. The two-to-one vote gave the industry hope that anti-GMO feeling in Europe could be reversed.

When I traveled to Switzerland before the vote, the outcome was still in doubt. In Basel, at the headquarters of Novartis, microbiologist Arthur Einsele confided to me that even his daughter, Claudia, a theology student, had qualms about genetic engineering.

"Daddy, is what your company does ethical?" she asked him one day.

The vote in Switzerland was a test case. Would a country slam shut its doors to an entire technology? For the ballot initiative was broad: It aimed not only to ban genetically modified crops but also prohibit the breeding of livestock with altered DNA and deny companies rights to patents on newly engineered varieties of plants and animals.

For the biotech industry, the stakes were enormous; besides Novartis, Nestlé and Hoffman LaRoche claimed Switzerland as their headquarters. The Swiss decision would reverberate throughout Europe.

In the town of Munchenstein, Florianne Koechlin, one of the organizers of the referendum, contended that Novartis and Monsanto "are giving us food we don't want." Koechlin and her allies believed that they still could prevail. But after one glimpse of the industry's ad campaign, I knew the anti-GMO forces were about to get smoked.

I had seen this movie before—in 1990, in California, when advocacy groups engineered a ballot initiative called "Big Green." Big Green would have imposed restrictions on farm chemicals and made a host of other environmentally friendly changes in the lives of Californians. Orchestrating Big Green was then State Senator Tom Hayden who, ten years later, would lead the drive in the California Assembly to require labeling of genetically modified foods.

In California, at the Mill Valley home of Bill Graham, the rock 'n' roll promoter who would die in a helicopter crash, I'd listened to Hayden and the Green strategists outline their strategy to Grateful Dead band members and well-heeled donors. Polls showed them ahead. But industry intervened with an ad blitz. Big Green was crushed in a landslide vote.

In Switzerland, Big Green was playing again. Television advertising is illegal in Switzerland, but print ads are allowed, and what the industry rolled out was potent. I recall the beaming young woman cooling down from exercise and sipping mineral water. *"Dank gentechnik habe ich mehr labensqualitat,"* she says, which translates to "Thanks to gene technology, I have a higher quality of life." In another, a middle-aged man—who, we're told, survived cancer—proclaims, "Thanks to gene therapy, my life is again worth living." Then there's the gap-toothed little girl playing with her father. "Thanks to gene technology, Daddy survived a heart attack," she says.

Monsanto was wiser than its rivals about the power of advertising. To gear up for the French citizens' conference, Monsanto ran full-page ads in every major French newspaper for nine straight days trumpeting the benefits of biotechnology. The French campaign was coupled with a run of ads in Britain that would provoke a debate far beyond the company's expectations.

Monsanto also had contributed liberally to the Swiss ad blitz, at least $250,000. But Novartis and the other European-based life-science companies refused to take part in Monsanto's campaign. They did not trust Monsanto—and not only because Monsanto was a business rival. Monsanto's hard-charging style troubled the European corporations, prompting them to keep the company and its political maneuverings at arm's length.

Before France, Monsanto's attitude had shifted. Seeing an uprising on the horizon, the company was making public acts of contrition for not doing a better job of preparing the European public for the new technology. In the spring, the company bent to consumer demands and softened its opposition to labeling modified products. "We now see that Europe views labeling somewhat differently than people in the U.S. do," Monsanto's Tom McDermott told me in Brussels.

Meanwhile, Carlos Joly, another Monsanto official, proclaimed: "We've heard the reaction loud and clear. We will respond by being more open, more informative, and more proactive toward European public opinion."

In France, on the eve of the citizens' conference, Monsanto aimed to shift that public opinion. On Sunday, June 7, the company's first newspaper ads appeared. Privately, Monsanto officials had begun describing their campaign across the Atlantic as "the European wars." The warfare was fought on political ground as well as grass roots. European politicians—among them Corrinne Lepage, France's former environmental minister—complained publicly about the company's tactics.

Before the citizens' conference opened, Lepage, a member of the Conservative party, accused Monsanto in a column in the newspaper *Le Monde* of seeking to brainwash the French by means of "elaborate techniques of battle."

DAY THREE: CONSENSUS, CAVIAR

Only in Paris would a government news conference feature tables creaking with caviar and silver bowls chilling magnums of champagne.

The French citizens hadn't just worked a few hours as Le Deaut, the moderator, had suggested. They met through the night, until 8:30 A.M., arguing how to fashion their recommendations.

In the morning light, Le Deaut praises them and apologizes for the tense moments in the first two days. "They have had no fear to ask questions that were a bit direct sometimes," he says. "This is just the evolution of democracy."

In contrast to the tough questions they had shoved in the faces of industry representatives, the citizens' recommendations are decidedly middle-ground—at least by European standards.

The fourteen take turns reading their conclusions.

Mireille Roine, a retiree and, at fifty-eight, the oldest member, announces that the citizens group found no health threat to humans from genetically modified food. Even so, from now on, they want citizens involved in government decisions and studies conducted by scientists with no connection to biotechnology companies.

François Rey, a twenty-year-old political scientist and the group's youngest member, declares that the citizens believe that gene-altered crops pose little risk to the environment with one exception; the "marker genes" for antibiotic resistance built in to tell if the newly engineered genes take hold. The group wants marker genes banned.

Rey makes a point I regard as critical in the public's thinking about genetic engineering. "In the end, if it improves the tastes and qualities of vegetables, we might be in favor," he says. His comments reaffirm to me that the resistance I was seeing in Europe might be overcome if and when the industry succeeded in offering products with benefits to consumers.

Planche, the insurance inspector, says the panel decided French farmers may need transgenic crops to remain competitive. But they also reached a conclusion that the industry found distressing: They demanded changes in French law so that companies would be held responsible for damage from modified crops, such as contamination of nearby organic produce.

In the end, the French citizens split on a key decision: Some of them wanted a moratorium on transgenic crops, others didn't. So they did not recommend a ban. That is a victory for biotech companies, which can assert rightly that French citizens rejected the moratorium that critics demand.

During a break, I track down Stephane Pasteau, the Monsanto representative, who was shaken two days before by the antibiotech tone of the conference's first day. He is relaxed, smiling. "Pretty balanced," he says of the recommendations.

Amid toasts, France's citizens' conference is declared a success on two fronts: On the democracy front, the aim of such gatherings is promoting a better-informed citizenry. The fourteen French brought randomly to Paris sifted through a barrage of technical information and conflicting views. They understood both the science and the politics of biotechnology. On the genetic engineering front, the citizens showed that accord can be reached on one of the most consequential issues of the day. For, in the end, they agreed on many points.

Of course, democracy can take surprising turns. Georges Schirm, the farmer who talked of duty to country, left disgusted. "It reinforced my skepticism in matters of how political decisions get made," he tells me.

And Claire Falhon, who began as the firebrand of the group, in the end argued behind closed doors against the moratorium, she confides to me afterward.

The great French citizens' conference is over, but the debate on biotechnology is not. What happened in Paris, Falhon remarks as she and her cadre of citizen policy makers head for home, "is only the beginning of a great debate."

PLANTINGS SEVEN

Last week, my wife, Sandra, looked at my genetically engineered soybeans for the first time. Her hands were on her hips.

When I told her a few weeks ago what I was up to, she
gave me the fish eye but didn't say anything. Today, she did.

"I don't know if I like the idea of having this growing here
and you squirting chemicals all over the yard," she said. Our
dog, a huge old yellow Lab named Max, was standing beside
her, looking at me disapprovingly.

I shook my head. I told her I would pull the weeds from
now on rather than spraying them but that the plants were part
of "an important scientific experiment."

Early this evening, I saw her standing by my soybean patch
again. So were neighbors, a couple who had dipped down into
my yard during the evening walk. Perhaps they were observing
the new blooms on the magnolia tree.

Later, Sandra asks me how long before I harvest the plants.

"They're not big enough. They're barely teenagers," I say.

"You know, your DNA specials aren't the favorite crop in
the neighborhood," she tells me.

14

IN BRITAIN, "ABSOLUTELY UNSTOPPABLE"

Genetic engineering *takes mankind into realms that belong*
to God and God alone.
—Prince Charles

It is inadvertently affirmed to the Christian communities of Europe
that the English are fools and madmen.
—Voltaire

THE THEME FROM *Mission Impossible* wafts through the air as protesters in
ghostly white uniforms converge on the perimeter of the Model Farm.
They gather slowly, their identities obscured by the hoods of their biohazard suits. Their only sound is a rising trill, muffled by their surgical
masks, which also are white. It occurs to me that they have copied this
high-pitched warbling from the warriors in *Zulu,* the film featuring their
countryman, Michael Caine, as a British officer guarding an outmanned
garrison in southern Africa. Only today, in the Oxfordshire town of Watlington, the white British, not the Zulus, mass for attack.

A helicopter hangs over the Model Farm's experimental field of yellow-
flowering oilseed rape plants, the focus of all this attention. In a grove of
trees, eighty helmeted British bobbies take the shade, waiting for orders
but, like me, unsure what will happen next.

Never have so many people, five hundred or so, massed to threaten
a genetically engineered crop. Never has a band of agricultural saboteurs
gathered so brazenly, in plain view of police and a forest of cameras:

television cameras, newspaper cameras, police cameras. On a stake, a sign written in typical British understatement, reads: *Do not disturb any of the plants growing on this site. Some of the plants are protected by a court injunction, and if you interfere with them in any way you may be in contempt of court.* Its wording is so milquetoast as to seem a dare rather than a warning.

But sabotage is not to be on the program this Sunday afternoon. What is—what has brought me from the United States to the English country-side—is a national anti-GMO gathering called the Stop the Crop Rally. Here, I expect to meet the activists leading the British wing of Europe's cresting anti-GMO crusade.

Many in the crowd are twentysomethings partial to tattoos and body piercings. Over their paper suits, dozens wear white tank tops stenciled "Greenpeace" in green letters. A few of them have strapped to their shoulders black-and-yellow paper wings, making them human monarch butterflies. They seem more a gathering of picnickers than protesters.

Among them, I'm told, are true underground saboteurs, the radical campaigners responsible for attacks throughout Europe. I have a list compiled by the biotech industry of thirty-four instances of sabotage, most on smaller, experimental fields of rapeseed. A veteran saboteur told me he knew of more than fifty sites trampled by anti-GMO raiders, usually in the dead of night.

By the light of this Sunday, I expect there'll be talk only. The rally, I'm told by a news release from an alliance called Genetic Engineering Network, "will be followed by a visit and walk-around the GM farm scale trial site." That means, I presume, that when the speechmaking ends, the protesters will march on the roads surrounding Model Farm. Then they will board their buses and go home. And I will head to London to continue my search for the roots of the rebellion I am witnessing today.

In Britain, epicenter of the European resistance to GMOs, a protest alongside a canola field in July 1999 evolved into the biggest sabotage of a modified crop to date.

SHIFTING FORTUNES

I have watched from distant points as the storm gathered in Europe. Everywhere I traveled—in North America, South America, even India— the winds of resistance were blowing in from Europe.

A year earlier, in the first months of 1998, the biotech industry seemed about to finesse a swift and bloodless revolution in the industrialized world's food supply. The United States was conquered; Europe was yield- ing. Three harvest seasons after the first genetically modified seeds were sown commercially in the United States, nearly one thousand food items containing modified soya occupied grocery shelves in Europe. By allowing concessions on one front—a voluntary regime of labeling modified food— GMO retailers contained their nascent European opposition.

Once modified seed was planted commercially on European soil, there would be no turning back. But in a matter of months, the balance shifted. In the brief but stormy history of genetically modified food, late 1998 to early 1999 was the period and Britain the country where counterinsur- gency rose.

The effect shook the biotech industry. Between February and May of 1999, Britain's major food retailers—Sainsbury, Safeway, Tesco, Marks & Spencer, and Somerfield—declared they would remove products with ge- netically modified ingredients from their shelves. McDonald's and fast- food outlets followed. Industry reps returning to the United States from Britain during these months spoke as though they'd fled the German bombs of World War II.

But I had not sounded the true measure of feeling and resolve until I returned to Europe in the summer of 1999.

In the months before, Greenpeace, Friends of the Earth, and an alliance of advocates had sown doubts across the continent. Saturating press cov- erage savaged the industry and the technology, and a new and potent biopolitics stemmed scientific advance.

The European Union benchmark approval sixteen months earlier of new genetically engineered crops—the Monsanto and Novartis hybrids of insect-resistant corn and AgrEvo's herbicide-tolerant corn and canola—

had been held up. As the tide of public opinion rose, the Union had not dared consider new applications. Meanwhile, France's High Court had blocked the second year of Novartis's corn planting after the first crop was decimated by a raid on the company's seed storehouse in the southern part of the country.

The advance of genetically modified food had stalled, and perhaps not just in Europe. In the United States, farmers listened nervously to reports from Europe and wondered if more markets for their modified crops would disappear.

In Britain, epicenter of the revolt, Mad Cow Disease had provided fertile soil for the seeds of opposition. In the summer of 1999, newspapers unraveled the tangle of blunders that had, in the 1980s, led to the epidemic of bovine spongiform encephalopathy, or B.S.E., the brain malady in cattle that had killed at least fifty people, caused the slaughter of over four million cattle, and cost billions of dollars.

Investigations revealed that then–prime minister Margaret Thatcher and her cabinet had dismissed early warnings about potential problems. Her industry-friendly government had relaxed rules regulating animal feed. Hence fewer precautions—such as sufficient heat to kill bacteria—had been taken when the remains of slaughtered animals were processed into livestock feed.

The cost of the epidemic was measured in more than lives and money: It cost the government and its scientists the faith of its people. Lost faith in the system that underwrites the nation's food safety nourished hostility to genetic engineering. It was bad luck for Monsanto and its rivals that the Mad Cow reckoning coincided with their campaign to plant European fields with modified seeds. They still might have neutralized that temporal coincidence had they won over the people before pushing their technology to market. Now they were paying the price.

But, as I was beginning to understand, the roots of the anti-GMO sentiments ran deeper. How deep, Americans were unwilling or unprepared to understand. For U.S. politicians, the whole fuss was a trade debate. To them, those Europeans were protectionists, wrongheadedly turning away an efficient farm technology. There is, in fact, a kernel of truth in this. But there's much more to the story.

When it comes to what they eat, Europeans are not Americans. On the one hand, famine and privation remain fresh in their memories. Threaten their food, and you threaten their survival. On the other, traditions go centuries deeper, and growing, preparing, and enjoying food remain active traditions in many households. Throughout much of Europe, food is a daily sacrament. Yet another force, the "Slow Cities" movement, was fomenting its own holy war against globalization, genetic engineering, and fast-food outlets.

"McDonald's never would have gotten off the ground in Europe," a United States Agriculture Department official told me—though it flourishes there today.

Also working against GMOs are the full bellies of modern Europe, which has all the food it needs and more. So why hurry to tinker with its genetic code, making changes that might never be undone?

The year 1999 had seen the largest global plantings ever of modified crops: nearly seventy-three million acres in the United States and another twenty-six million in the rest of the world. Would 1999 be just a beginning? Or would it be the peak of a short-lived technological revolution?

WAR OF THE WORLDS

Most of the protesters have come to Watlington by bus, from London, Manchester, Totnes, Swindon, Aylesbury, and nearby Oxford. In the meadow where they legally rally, the placards are more artful and certainly more forceful than the mealy-mouthed warnings posted by the government. Along a fence, they have stretched orange banners that read in blue letters *Resistance is Fertile*.

I see signs advising "Caution: Pollen from GenCrops Contaminates This Area." The only visible contamination I can find in the meadow where they have gathered across from the experimental field is cow manure, some of it fresh. To my surprise, I also find people like W.E. Stonelake, who is neither young, tattooed, nor pierced.

Stonelake, seventy-nine, a shop steward in a power plant during his working years, looks the essence of English in his tweedy coat and bowler.

When I ask whether I might speak with him, he surprises me with a question of his own. "Have you read H.G. Wells's *War of the Worlds*?" he asks. His literary allusion to an invasion of Martian techno-monsters suggests a paranoia about genetically engineering our daily bread.

"When you start something, you never know what the consequences are going to be. It could bring changes we haven't dreamed of. Things in the world change naturally. And if we are going to change in harmony with them, fine. But that isn't happening," he tells me.

But, I ask him, what about the good we are sacrificing if we stop genetic research. "I don't think it will feed the poor of the world," he responds. "More food has to come from the heart of man, not from chemical operations."

While writing down his words, I hear Alan Simpson, the Member of Parliament, inveighing against twenty-first-century global imperialism of multinationals, or something to that effect. I am reminded that GMOs have become an issue in parliament, while in the American Congress, they are barely discussed.

I prefer talking with rally-goers, not politicians and celebrities, until I hear the amplified, fetching voice of Lynda Brown, a food writer talking about food. "Millions of Americans are eating mutant potatoes, which are classified as a pesticide. Doesn't that make you go 'yum, yum'?" she says.

"The real question," she continues, "is do we want to eat them? Or do we have the right not to eat them?"

When I replay my tape of her speech, I hear predictions of a future with "GM warfare, GM forests, a diet riddled with GM foods, and a GM landscape . . . Do we continue down the industrialized agriculture path and forfeit our right to the kind of food we want to eat and the kind of environment we want to live in?" she asks. "Or do we stand up for our rights as human beings and consumers and choose to support a sustainable agricultural path which will truly give us food we can trust, which will truly safeguard our environment, and does offer the best hope for our planet? That's why I'm standing in this field."

When I catch up to Brown, a well-tanned blonde, she tells me that she lives just five miles from where we stand. She's part of an anti-GMO campaign by food writers, some of them from leading daily newspapers, organized by Greenpeace. Their public statement, also signed by chefs, reads: "As food professionals, we object to the introduction of GM foods into the food chain. This is imposing a genetic experiment on the public, which could have unpredictable and irreversible adverse consequences."

Brown is the author of *The Shopper's Guide to Organic Food* and other books. She and Darina Allen, the television chef who runs the Ballymaloe Cookery School in Ireland, are among prominent food experts turning European public opinion against the products of genetic engineering.

How, I ask her, can British journalists, especially food writers, discard their objectivity when it comes to modified food.

"You have bigots on both sides of the fence, journalists who are absolutely rabid who are not prepared to look at both sides of the issue. And quite frankly, it's good copy," Brown says.

AFTER "MAD COW"

For years, the British public suspected that their political leaders, their scientists—indeed, the country's whole regulatory apparatus—failed them by not arresting the spread of Mad Cow Disease. But only recently had the evidence confirmed those suspicions. Urged by the British Association for the Advancement of Science, the Blair government investigated the handling of the epidemic.

During my visit, a headline in the respected *Financial Times* told a critical part of the tale: "Thatcher 'feared ban would damage interests of food producers,' " the headline read. The British public devoured the unfolding story as if it were a serial novel. Each day's newspapers published new revelations of complicity.

The Mad Cow epidemic began in 1984, in West Sussex, in southern England, where a dairy cow identified as No. 133 displayed what veterinarian David Bee described as "a variety of unusual clinical manifestations." Two months later, No. 133 was dead, and more cows on the farm lost their coordination and their placid dispositions.

Was it poisoning from toxics, perhaps mercury or lead? Government scientists had no inkling. But after another half dozen West Sussex cows died, a stricken animal, No. 142, was slaughtered for autopsy. What pathologists found were spongelike holes in its brain. The diagnosis: spongiform encephalopathy.

Not until four years later did epidemiologists pinpoint a cause—a diet of feed enhanced with protein from the ground-up remains of slaughtered animals. Dr. Daniel C. Gajdusek, the Nobel laureate researcher, referred to this method of livestock feeding as "high-tech cannibalism."

Scientists remain puzzled about how the disease works. It is not believed to be caused by bacteria, a virus, or a fungus. Rather, the culprits are thought to be infectious particles called prions that, working with a virus or other agent, kill cells and riddle the brain with holes.

Only in 1989 did the government ban the use of the brain, thymus, and other bovine offal in feed. By then, people worried if the disease could be transmitted to humans.

Then, in the summer of 1994, Stephen Churchill, a Royal Air Force cadet, became strangely ill. Doctors considered Creutzfeldt-Jakob disease, a neurodegenerative disorder that resembled B.E.S. but usually strikes elderly people. Churchill died in May 1995 and by fall, other young Brits began to show symptoms.

Still, the government insisted that beef products were safe. "There is no conceivable risk from what is now in the food chain," then–health secretary Stephen Dorrell declared in early 1996.

Just four months later, in April, Dorrell was eating his words in front of the House of Commons. Ten people had died, he announced, and the most likely explanation was "exposure to B.S.E." The British were shell-shocked to learn of a new disease—a new variant Creutzfeldt-Jakob disease—referred to as human B.S.E.

The public was in for another shock: Two of the victims had been blood donors. To reduce the potential for spread of the disease, Britain began importing much of its blood for plasma and established a seventy-seven-million-dollar program to remove from donated blood the white blood cells that might carry the B.S.E. agent.

The country stewed over how many more cases of B.S.E. would surface—over how many years. Britain's scientific establishment, beleaguered already, did not assuage national fears. They confessed they didn't know how much infected meat caused the disease or the length of its incubation period. Perhaps only a few people might succumb; perhaps tens of thousands could fall victim.

Britain's Mad Cow outbreak spread across Europe, raising suspicion of the regulatory bodies whose imprimatur the biotech companies sought. In the late 1990s, public response behaved like a disease itself, receding at times and then spreading, finding new niches in which to take root. By early 2001, fear of the brain-wasting disease was deeply rooted on the continent. In Germany, both the health and agriculture ministers were driven to resign in a scandal over lax testing prior to a German outbreak. In France, farmers cast stones at their prime minister, and police raided French government offices to investigate leveling manslaughter charges against bureaucrats for ineffectively blockading the disease. In Spain, cattle ranchers blockaded slaughterhouses to protest what they viewed as draconian emergency precautions imposed by the government on meat production. The European Union agreed to spend nearly one billion dollars to help bail out farm businesses clobbered by a combination of new production costs and a 30 percent plunge in the price of beef.

Two months before the Watlington protest, yet another food scandal erupted in Europe with the disclosure that Belgian farm animals were fed with dioxin-contaminated meal. As a result, Belgian chicken, eggs, beef, and pork were recalled from the European Union market—and the Christian Democratic government of Jean-Luc Dehaene fell.

Once again, where were the regulators whose duty is to protect people?

AN "ALTERNATIVE MODEL FARM"

Nine days before the Watlington protest, at two hours before dawn, twenty protesters moved into a farmhouse that sits in a grove across the road from the test field of rapeseed. The house, like the field, is owned by the Ninth Earl of Macclesfield who, over the years, has been notably averse to trespassers.

As well as strife over gene-altered food, class warfare has flared in Oxfordshire. Lord Macclesfield, who is in his late fifties, lives in a moated castle on three thousand five hundred ancestral acres. He is descended from the amateur astronomer, Lord Macclesfield, who was president of the Britain's most prestigious science organization, the Royal Society, in the late eighteenth century.

For much of the 1990s, Lord Macclesfield fought against people who demanded the right to cross his land. Country walkers proliferate in Britain to a degree rural dwellers in the United States aren't accustomed to, and Lord Macclesfield's five hundred acres of uncultivated land were an alluring destination to folks hankering for a stroll. When the Ramblers Association wanted parliament to legitimize their rights to move about the countryside, Lord Macclesfield became a national symbol of the propro-perty forces. "What are they so frightened of? All ramblers do is carry cheese and tomato sandwiches," a member of the association joked.

Now, the anti-GMO forces not only had invaded Lord Macclesfield's land, they had sown organic seed there and renamed the property Alternative Model Farm. His Lordship was not happy.

"It is a disgrace these people should occupy private property," he said. "The place to make their protest is at the Commons, where they would be locked up for their pains. Unless we give proper trials to these crops, as Tony Blair intends, we will not be able to see if they are safe or not."

"ONGOING COLLAPSE"

By early 1999, the debate over genetically modified foods was as much a part of British culture as royalty and rock 'n' roll. Anti-GMO activists demonstrated heavy clout when, in 1998, they lured Prince Charles to their cause.

Writing in the *Daily Telegraph,* the Prince of Wales made his startling declaration that scientists have strayed into "realms that belong to God and God alone . . . We simply do not know the long-term consequences for human health and the wider environment of releasing plants bred in this way . . . The lesson of B.S.E., and other entirely manmade disasters on the road to 'cheap food' is surely that it is the unforeseen consequences which present the greatest cause for concern."

In his op-ed piece, Charles recalled the disasters of overreliance on a single variety of a crop. "Yet this is surely what genetic modification will encourage," he wrote. "It is entirely possible that within ten years, all the world's production of staple crops, such as soya, maize, wheat and rice, will be bred from a few genetically modified varieties—unless pressure dictates otherwise."

Charles scoffed at biotechnology's promise of feeding more people. "Will the companies controlling these techniques ever be able to achieve what they would regard as a sufficient return from selling their products to the world's poorest people?" he asked.

Charles's salvo hit home. Industry officials were apoplectic at the prince's pronouncements, accusing him of a conflict of interest because of his organic farming operation. In some circles, he became known as "the Luddite Prince." But some of Britain's revered figures followed suit, among them Sir Paul McCartney, the former Beatle, who vowed to spend millions of dollars to make sure that the products of his late wife's company, Linda McCartney Foods, did not include modified ingredients.

The GMOs issue penetrated deeply into the culture. *The Archers,* a radio soap opera on the BBC, ran a series of episodes depicting the debate. In one, a character was punched when he confronted protesters attacking a field of genetically modified crops. Environmentalists complained to BBC that none of the incidents of sabotage had involved violence to humans.

Anti-GMO campaigners assaulted not only British fields but also British minds. In January, they began distributing what they called the Natural Reality SuperWeed Kit 1.0. They insisted that their mixture of naturally occurring and genetically mutated seeds would, if allowed to germinate, create a "Superweed" resistant to Monsanto's Roundup and other herbicides. "If released, SuperWeed 1.0 will not only destroy the profitability of all GM crops, but also of conventional and organic crops. This genetic contamination will be irreversible," the activists said in their news release. Their point: Genetic manipulation can yield scary results.

Pro-biotech forces had been fond of branding detractors as part of a radical fringe. But such labeling lost credibility when the centrist Consumers' Association, English Nature, the Soil Association, and the Vegetarian Society denounced the technology.

The rush by supermarkets to banish modified products left the biotech industry reeling. Competitive pressure had commenced when Iceland Foods, the first food retailer to ban GMOs, trumpeted that its sales had soared by 12 percent. In newspaper ads that were extraordinary by United States standards, Iceland declared in the fall of 1998 that "the U.S. president doesn't care what you put in your mouth." The ad asserted that Iceland banned genetically modified ingredients "because we refuse to produce food that we wouldn't be happy for our children to eat. And we did it because we trust our customers' instincts more than those of the food industry."

By July of 1999, every major retailer had followed Iceland's lead. On the day that Britain's second-largest food distributor, Sainsbury, announced that it had succeeded in eliminating gene-altered ingredients from every Sainsbury-labeled product in its 415 stores, I asked the company's Gillian Bridger about her company's motivation. "Our customers were telling us, 'we just don't want it in there at all.' So we decided to go whole hog and remove it all," she told me.

Restaurants also hastened to take advantage of the inflamed public opinion. By the end of 1998, nineteen of Britain's top twenty-three restaurants as rated in *The Good Food Guide 1999* had signed on to a Friends of the Earth initiative that called for a five-year ban on genetically modified foods. The authoritative *Guide* ran an editorial asserting: "To introduce experimental, herbicide-resistant crops without some soundly based assurance is madness, albeit perfectly legal madness . . . If B.S.E. has taught us anything, it is surely to be cautious about tampering with natural processes, however well-intentioned, however plausibly the benefits are packaged."

In July of 1999, I set out in London to track worried chefs. I ended up at a table in The Square, on Bruton Street, which I've seen characterized as a home for "power-lunching captains of industry." While I look at modern art on the mauve-colored walls, a tuxedoed waiter presents me a wine list that includes a bottle of '45 Château Mouton Rothschild for $8,400. British cuisine, I learn swiftly, is not strictly bland shepherd's pies, fish and chips, and cold toast smeared with Bovril. I choose an appetizer of duck foie gras and an entrée of thinly sliced rump veal with borlotti beans. This is lunch.

Chef and owner Philip Howard, who was born in South Africa and studied microbiology at the University of Kent in Britain, is known around town for his exacting pursuit of the finest ingredients. The unassuming Howard, who is in his early thirties, tells me that he is aware of the potential of genetic engineering to one day improve food. But he is troubled, he says, that neither the genetic engineers nor the breeders who deploy traditional means pay attention to the tastes of their creations. "They want crops that yield better, change the color of food, and the like. But nothing is done for taste," he says.

In 1999, the campaign reached beyond human food to what British livestock eat. News reports in the spring told of a plan by supermarkets to stop selling meat fed with grain that was genetically modified. I was certain that the campaign had reached its limits when, during the time of my visit, advocacy groups demanded that London department stores identify clothing made from genetically engineered cotton. Stores responded that there's no test to tell.

The debate starts young. At a kids' parliament in the city of Birmingham in the spring, children ten and eleven had debated this loaded proposition: "Genetic engineering, which includes animal cloning and genetically modified foods, is the biggest threat to mankind since the advent of nuclear weapons."

The depth of anti-GMO sentiments in Britain showed up in polling done for Monsanto by Stanley Greenberg. Greenberg had handled survey research for Bill Clinton's successful presidential campaign in 1992 and for the White House after that. But he left his party's employ after their 1994 off-year election debacle in which Republicans captured the Congress for the first time in nearly a half century.

Greenberg was a Democratic pollster with a track record and plenty of experience in Britain. And Monsanto was a company known for its alliances with Democrats. So when the company needed to take stock of the damage in Europe, they knocked on his Capitol Hill door.

What Greenberg found in October of 1998 couldn't have been imagined by company officials just a year before as they stood boldly at Europe's garden gate preparing to sow millions of modified seeds. Leaked to Greenpeace, his findings reinforced the widening view that not just Monsanto but the entire biotechnology industry was in trouble.

Greenberg wrote: "The latest survey shows an on-going collapse of public support for biotechnology and GM foods. At each point in this project, we keep thinking that we have reached the low point and that public thinking will stabilize, but we apparently have not reached that point.

"The latest survey shows a steady decline over the year, which may have accelerated in the most recent period . . . Overall feelings toward foods with genetically modified ingredients have grown dramatically more negative, which is probably the best measure of our declining fortunes in Britain. Only about twelve percent have reacted positively over the last year, but negative feelings have risen from thirty-eight percent a year ago to forty-four percent in May to fifty-one percent today. A third of the public is now extremely negative, up twenty percent."

In Britain, Greenberg found a public cool not just to genetic engineering but to scientific progress in general. To one of his questions, two-thirds said they were skeptical that science would benefit the ordinary person. The polling also found confidence slipping in Britain's regulatory agencies, which Greenberg blamed on Mad Cow. His interviews showed that Monsanto had become an issue in itself, and part of the reason, Greenberg said, was the "worsening and deepening" feelings in the media.

"The media elites are strongly hostile to biotechnology and Monsanto," he wrote. "They think the government is being too lax and believe they must expose the dangers."

But the media were just part of it, as Greenberg conceded. "There are clearly large forces at work that are making public acceptance in Britain problematic," he wrote.

RAPE TO CANOLA

Rapeseed has been cultivated for twenty thousand years; references to it appear in the earliest writings of Asian cultures. In the thirteenth century, the wild cabbage came into wide cultivation in Europe because it tolerated cooler climates better than other oilseed plants. In the pods that form on the thigh-high bushes is an oil that lighted many lamps and cooked centuries of meals for the ancestors of the legion of protesters poised alongside the Model Farm.

After the Industrial Revolution, it was discovered that rape also yields oil suitable for lubrication. That's when the North Americans became interested in the plant. In the second half of the nineteenth century, with whale-oil prices climbing, American farmers were encouraged to grow canola to fuel the nation's lighthouses. In the United States and Canada, the genetic derivation of rapeseed goes by the name canola. In 1985, the Food and Drug Administration accorded the status of "Generally Recognized as Safe" to canola oil, and its use as a cooking oil has increased since.

In my kitchen cupboard in Maryland is a plastic bottle of canola oil that comes from a whole foods supermarket. I use it to cook popcorn. The label tells me that it was produced in Canada and that it is 93 percent free of saturated fat. Because I live in the United States, where labeling of gene-altered foods is not required, it does not tell me if my oil comes from the tens of thousands of acres in Canada sown with genetically engineered Roundup Ready canola seeds.

In Watlington, the field that has drawn protesters from across England (and me from Washington) is green with rapeseed genetically modified to resist AgrEvo's Liberty herbicide. In other words, the farmer can spray the weed killer directly on the plants; the rapeseed plants will remain healthy, but weeds will die. The field test in Watlington, one of six farm-scale trial sites being run by the government in the United Kingdom, is designed to measure the effects of the modified crops on the abundance as well as the diversity of plants and invertebrates.

The pods begin to form at the bottom of the bush and continue to grow as the rape produces its yellow flowers. The pods on the plants look like green beans of four or five inches in length. Inside, the pod is divided into halves by a membrane. Together, the compartments contain about twenty-five pea-sized seeds. In pods near the top, seeds are translucent like tiny bubbles; near the bottom, they are green and turning to black, a sign of maturity.

These plants have expended a lot of energy growing from seeds. But now, according to whispers from protesters, they may die before harvest.

"THIS TERRIFYING TAMPERING"

The British news media differs from it counterparts in the United States in fundamental ways, which is part of the reason for the hostility that pollster Greenberg found. In Britain, papers are sold primarily at news-stands rather than by home delivery, so they compete for readers with headlines, "the wood," as the tabloids like to say. In Britain's thriving tabloid industry, rules for fairness are much looser than in the United States. Yet another differentiating factor is television; in Britain, television isn't anywhere near as important in delivering news to people as in the United States. Because of its nuances and shortages of ready visuals, the debate over genetic engineering makes challenging coverage for television.

In early 1999, British tabloids laid the wood to the biotech industry with these headlines:

> MUTANT CROPS COULD KILL YOU.
> HUMAN GENES IN GM FOOD.
> THIS TERRIFYING TAMPERING.
> COURT RAP FOR FRANKENSTEIN FOOD FIRMS.
> BLAIR MONSTERED ON FRANKENSTEIN FOODS.
> FAST FOOD GIANTS BIN MUTANT GRUB.

In the month of February, a Monsanto European official counted 712 newspaper stories, many of them in tabloids, about the genetic engineer-ing of crops. The *Daily Mail* pronounced its coverage a campaign and ac-companied stories, with a logo of a puzzle legended "Genetic Food Watch." The *Express* had a logo of its own with the words "Safe Food." Even the venerable *Times of London,* in a February 22 story about a ge-netically engineered dairy hormone, asked, "Is Frankenstein's milk around the corner?"

A February 17 story in the *Guardian* business section opened with this question that parodied the overheated prose: "Will Frankenstein foods cause two-headed rabbits to sprout in fields otherwise denuded of life except for giant tomatoes?"

Often, stories matched the sensational headlines. On January 30, a *Daily Mail* article with the headline "Can Frankenstein Foods Harm Your Unborn Baby" opened with this paragraph: "Health experts investigating the impact of so-called Frankenstein foods have suggested examining abortion records."

Five days later, a *Guardian* story began: "An outbreak of a fatal disease that infected 5,000 people, killing 37 and leaving 1,500 permanently ill, was linked to genetically modified food, a Labor MP claimed in a Commons debate yesterday." The article referred to an outbreak of sickness and death caused by a bacterium—not food—used to make a vitamin supplement. And while DNA had been manipulated, it was not known whether that played any role in the bad batch of vitamins.

A *Daily Mail* article on February 6 kept pace: "Disturbing questions about the government's policy on so-called Frankenstein food were raised last night when it emerged that a producer of genetically modified crops has given money to the Labor Party." The story went on to allege that Novartis had helped to bankroll a seminar in which politicians took part.

Liberal papers, too, shined the harsh light on Tony Blair. "For a man who has always seemed to understand the issues that really matter to the public," a *Guardian* editorial began, "Tony Blair's touch appears to have abandoned him on genetically modified food."

In May, I found myself on the fringes of a critical story when I showed up at the Meridian Institute, a Washington think tank that receives United States government funds to put on seminars for foreign visitors, to speak to a group of British journalists, farm experts, and scientists. (I received no money for my participation.) Under a *Daily Mail* headline "Anger at propaganda freebie," the story began: "A row has erupted over a propaganda campaign set up by the U.S. government and genetically modified food giant Monsanto." The story went on to say that before touring Monsanto headquarters in St. Louis, "the ten delegates will be wined, dined and bombarded with positive messages about the safety of genetically modified crops and food."

There was more at work than competition and, even by tabloid standards, sloppiness. For several years, the Tories had watched their party pilloried for the Thatcher government's bumbling during the Mad Cow outbreak. Toward the end of the 1990s, the opportunity for payback arrived in bags of genetically modified seeds unloaded at British docks. In 1999, the Tories and their supporters in England's partisan press seldom missed the chance to pillory Tony Blair for linking arms with U.S. President Bill Clinton to support genetically modified food. Blair and genetic modified food proved to be a dependable combo for politicians and reporters alike, as did Monsanto.

"Public Enemy Number One," was how Norman Baker, a Conservative in parliament, referred to Monsanto during a debate in March in the House of Commons.

Blair and Cabinet Officer Jack Cunningham, whom biotechnology companies saw as their closest ally in the government, withstood the heat better than Monsanto, branding the press accounts "scare stories." But just weeks before the protesters gathered in Oxfordshire, Blair had conceded in a widely publicized interview that "the jury is out" on the safety of modified food.

With his boyish looks, Tony Blair is a handsome man, but no longer can I look at him without remembering his scary countenance as depicted in the *Mirror* in February 1999, with his huge forehead with shocks of hair covering a scar. **THE PRIME MONSTER,** not the prime minister, the fat, black headline screamed. Decked beneath were the words "Fury as Blair says: 'I eat Frankenstein food and it's safe.' "

DOWN DEEP, A MONSTER LIVES

The biotech barons blame environmentalists who, they say, frighten people to raise money to keep themselves in business. They blame the organic industry for exploiting uncertainties for profit. They blame the news media for telling a side of the story that is not theirs. Occasionally, they blame one another, though seldom do they blame themselves.

Perhaps they should blame Mary Shelley, the author of *Frankenstein,* the 1818 classic.

I am convinced that the miserable, exploited, and nameless monster that the world knows as Frankenstein is a player in the biotech wars. He may lie dead on some ice patch in the frozen North, but he survives in the minds of millions, and he lumbers about at the heart of Monsanto's misery.

"Frankenstein food" and "Frankenfood" have, in a few short years, come to represent all that is unknown, frightening, and, indeed, monstrous about a technology that manipulates life. They are bywords in a debate not just about the safety of science but also its ethics. There's more to the Frankenfood follies than a cute name.

It was the summer 1816 when the poet Lord Byron suggested to fellow poet Percy Bysshe Shelley and his eighteen-year-old wife, Mary, that they engage in a ghost-writing competition. The Shelleys were summering in Switzerland near Lord Byron, and with the unremitting rains that season, this trio of brainy writers needed a creative outlet; a game, if you will, to fill their time. The early nineteenth century was the era of eerie Gothic novels, directed toward the popular reader, and it also was a time of widespread fascination with animism—the belief that something inanimate could be brought to life.

Percy Shelley, besides being an ill-fated poet (he drowned in 1822 sailing in Italy), was an amateur scientist fascinated by chemistry. Mary Shelley, the precocious, London-born daughter of two writers, was drawn herself to the wonders of science; in her introduction to *Frankenstein,* she recalls an experiment of the day conducted by scientist Erasmus Darwin, who was said to have preserved a piece of vermicelli in a glass case "till by some extraordinary means it began to move with voluntary motion."

Throughout the tale of the creation of the monster, his rampages, his mysteriously learned musings, and, ultimately his meetings with his maker, Victor Frankenstein, Shelley imparts a message of the evils of technology: the dehumanizing potential of science performed by the "dogmatic experimentalists." In Shelley's introduction, she wrote that she wanted to produce a story that "would speak to the mysterious fears of our nature"—not unlike the fears that are greeting genetic engineers today.

Early in the book, Victor Frankenstein comes under the tutelage of Professor M. Waldman who, at a time when science was developing from two traditions, philosophy on the one hand and skilled crafts on the other, beckons his pupil with the allure of experimentation.

These philosophers, whose hands seem only made to dabble in dirt, and their eyes to pore over the microscope or crucible, have indeed performed miracles. They penetrate into the recesses of nature, and show how she works in her hiding places. They ascend into heavens; they discovered how the blood circulates and the nature of the air we breathe. They have acquired new and almost unlimited powers; they can command the thunders of heaven, mimic the earthquake and even mock the invisible world with its own shadows.

Victor Frankenstein, making pronouncements not unlike those of the genetic engineers of the late twentieth century, exclaims that he intends to "pioneer a new way, explore unknown powers and unfold to the world the deepest mysteries of creation." In *Frankenstein,* the reader is propelled not just by what science is doing now, but what science *might* do in the future, just as are those today who fear not what the geneticists have produced so far but the interspecies creations to come—the tomatoes with fish genes, the pigs with human genes.

Mary Shelly's *Frankenstein* is about more than just a creature; it's about a belief system, an implicit fear of technology that, by virtue of literature and film, has resonated in the psyches of hundreds of millions of people. And nowhere is *Frankenstein* more internalized than in the British.

I wrote while on an earlier trip to Europe that the term "Frankenfood" never would find its way into the stories of American journalists. I was wrong; when the GMOs debate moved westward across the Atlantic, Frankenstein and all that he conveys was ready to perform anew.

We know the fate of Frankenstein's monster; the fate of genetically modified food remains uncertain. The transnationals bringing the world genetic engineering need to hope that the story doesn't turn out like the book, in which the creation torments his creator even to death.

The Monsantos of the world, bolstered by their pollsters and their public-relations advisors, would never have imagined that the nightmarish creation in a book approaching two hundred years old could rise up to spoil their dreams. Of course, these companies probably also didn't know another little fact the monster disclosed to his creator when finally, in Chamounix, he tracks him down: He is a vegetarian.

"UNSTOPPABLE, ABSOLUTELY UNSTOPPABLE"

Lord Macclesfield would have felt more at home than I in the
Reform Club in London, where I stop on my way to the Watling
protest. If, that is, that famous conservative could abide its liberal tradition.
Sir William Asscher is waiting for me, settled with his gin-and-tonic into
a red-leather wing chair in the Audience Room where, in *Around the World
in Eighty Days,* Philias Fogg made his grand bet.

"If the darn demonstrators would let us sort it out, we would have the
answers in a few years. I hope to goodness that this will not continue,"
says Asscher, who is chairman of the British Medical Society's board of
science and education.

Two months earlier, Asscher's organization poured fuel on the fire by
calling for a moratorium on modified crops until scientists conduct more
safety studies. The association, which represents over 80 percent of Britain's doctors, recommended that engineered crops be processed separately
from conventional crops and that food with gene-altered ingredients be
labeled. These are the prescriptions that the American biotechnology industry and its allies in the farm fields and food business do not want filled.

The physicians had identified no health risks to genetic engineering;
moreover, they restated their conclusion of five years earlier that the technology holds promise, including the potential of reducing allergens in
food. But they recommended advancing cautiously in dangerous and un-
charted territory, as reflected in Asscher's words at the time. "Once the
GM genie is out of the bottle, the impact on the environment is likely to
be irreversible. That is why the precautionary principle is so particularly
important on this issue. It is even more serious than the licensing of medicines which, if necessary, can be withdrawn," he said.

In the United States, where the American Medical Association has
raised no such concerns, the British report pricked the sensitive skins of
senators whose districts included swaths of genetically modified crops
already planted for export. For them, the issue was trade. "It is characteristic of the European Union to hide behind studies such as this in order
to maintain its protectionist trade policies," fumed then-senator John

Ashcroft, a Republican from Monsanto's home state of Missouri, and later the Attorney General of the United States in the Bush administration.

The Reform Club, established in 1836, has counted among its members the politicians Churchill and Gladstone and the literary giants H.G. Wells and Arthur Conan Doyle. Henry James used to reside here. With its dark wood, plush carpets, and fifteen-foot-high shelves of books, the place oozes Establishment. But GMO anxiety permeates even the stuffy air of Britain's Pall Mall gentleman's clubs.

"I'm pretty certain there will be effects on the environment," Asscher tells me. When I ask what proof he has, he points to the monarch butterfly study newly reported in May from Cornell University, which showed—under laboratory conditions only—that pollen from corn with the Bt gene killed monarch larvae. He further enlists research at the Scottish Crop Research Institute, reported a few months earlier, that potatoes genetically engineered for pest resistance damaged ladybird beetles, which are known as the "farmer's friend" because they feed on aphids.

Even more portentous than Asscher's critique of genetic engineering is his political analysis. "What's happening here is unstoppable, absolutely unstoppable," he tells me. "It has to do with the British bulldog mentality in which we must know everything for ourselves."

PUSZTAI'S POTATOES

With conflicting, often distorted information rushing at them from many sides, the British had to be confused. Negative reports resounded at such decibels that I doubt if people could hear the positive news. In the summer of 1999, scientists at Britain's John Innes Center had reported finding a gene in wheat that controls the height of plants. That doesn't sound like much, but it could help to feed people. The scientists believe that if the gene proves to be transferable, they could engineer rice and other plants to direct more energy toward producing grain and less toward stalks. The prestigious journal *Nature* did the industry a favor with the headline, "Stunted GM crop may help feed world."

When I spoke with one of the Innes scientists, Michael Bevan, he said the study had been received well. But, he added, "We realize we are up against a situation where people are against the technology."

One of the reports that helped swing public sentiments against the technology originated in the work of scientist Arpad Pusztai, a Hungarian immigrant working at the Rowett Research Institute in Scotland. On *The World in Action*—a television report that had aired on August 10, 1998—Pusztai declared that his experiments showed that modified potatoes had damaged the immune system of experimental rats. For 110 days, he had fed rats with potatoes modified with genes for lectin, a protein which offers protection in the field against aphids and other pests. Rats eating the harvest of those engineered plants were smaller and less immune to disease than those fed on unmodified potatoes, Pusztai said.

"We are assured that this [modified food] is absolutely safe and that no harm can come to us from eating it. But if you gave me the choice right now, I wouldn't eat it," Pusztai said. He further alleged that biotech companies were treating the public like "human guinea pigs."

Pusztai's findings were regarded briefly as a milestone in scientific efforts to gauge the safety of modified food, and they triggered a new outbreak of condemnation. Pusztai claimed that three days later Rowett forced him, at age sixty-eight, into retirement, saying that he had released "misleading information." But his case was far from over.

Pusztai claimed in media interviews that he was wronged, and Prince Charles came to his defense. The Royal Society, Britain's premier scientific organization, criticized his findings. But Pusztai enjoyed a measure of vindication in 1999 when *Lancet,* the prestigious British medical journal, published a peer-reviewed paper that Pusztai coauthored. The study reported that rats fed those same altered potatoes had experienced intestinal thickening, an adverse reaction. It raised the possibility that the damage was due not to the lectin but to the process of genetic engineering itself.

Once again, Pusztai was denounced by biotechnology companies and their allies in the scientific community. The Royal Society declared the study flawed, adding, "It is wrong to conclude that there are human health concerns with the process of genetic modification itself or even with the particular genes inserted into these GM potatoes."

Only more time and more studies will show if Arpad Pusztai's fears hold up—or generalize. Whether he was on to something or not, the headlines he made further indicted GMOs in the court of British public sentiment.

A SABOTAGE

The music has changed from the *Mission Impossible* theme to the lilting melodies of Pink Floyd's "Echoes." Someone has turned the twin, six-hundred-watt amplifiers suitable for a band concert toward the field of genetically altered rapeseed. At an opening along the perimeter of the field, protesters have planted a yellow sign with a black symbol and the word "Biohazard." A thin, shirtless man is putting the finishing touches on what looks to be a scarecrow standing ten feet high. With its white shirt and black necktie, it is unlike any scarecrow I have seen. A barefoot young man in white slacks, yellow sweatshirt, and a white Arab headdress dances wildly nearer and nearer the field.

The rally in the meadow is over, and the participants poise, like runners at the start of a marathon. As they pass through a wood-plank gate, some of them linger, gazing at the test field. Others march toward the road, where I am standing. Neither group is headed toward the buses that have delivered them from distant points. As if on cue, a contingent of about one hundred people steps across the ditch that separates the road from the test field. This group breaks in half; now there are three streams of marchers, two of them on the edge of forbidden property, one on the road. It is a diversionary tactic, I conclude. The police don't know what to do, and neither do I.

The protesters are in no hurry, but they advance purposefully, some of them carrying white flags with the biohazard symbol in red and the word "NO." Leading the procession along the rapeseed are two women wearing butterfly wings. Between them is stretched a twenty-foot-long yellow banner reading in black letters: "Stop GenetiX Crops!" Beneath the word "Crops" in smaller letters runs the word "Greenpeace."

A phalanx of the marchers turns into the field and commences what will become the biggest and most brazen farm sabotage in the short but tempestuous history of genetically modified food. The white-clad attackers no longer walk, they trample, and as they advance they yank rapeseed plants from the earth, flinging them in the air. Flying GMOs, I write in my notebook. I see people throw themselves to the ground, rolling, flattening and uprooting the plants with both hands. It is an unnerving, terrifying scene, accompanied by haunting rock music.

I have covered protests of all sorts during twenty-five years as a reporter. Almost always there is a pivotal moment when the course of events is decided, when the news of the day becomes determined. I recall one such moment, in New Jersey nearly a decade ago, when police unexpectedly rushed in, beating and handcuffing protesters who had lined up menacingly outside a chemical plant protesting toxic exports I had disclosed. On that day, I took a level, left-handed swing of a baton to the solar plexis from a SWAT team member, and since then I've kept my distance from police at demonstrations.

In Watlington, that critical moment has arrived. From their grove of trees, the bobbies march single file into the field. Where they walk, the rapeseed plants are flattened. Lord Macclesfield has arrived in a Range Rover. Crimson with anger, he stands by.

But this is England, not New Jersey. Or Iowa, where I've heard farmers say they've got shotguns ready for anti-GMO trespassers. Even with Lord Macclesfield looking on, the bobbies don't stop the saboteurs, let alone beat them. For more than an hour, the invaders tear away plants and stamp them flat, destroying a swath of the field and ruining the field test at the Model Farm. Six people are arrested, charged with criminal damage to property. Police don't explain why the half dozen are singled out.

Perhaps their tentativeness is rooted in the public antipathy toward GMOs in Britain. I am baffled by the motives on all sides. The protesters say genetically modified crops must be proved safe in field tests. But they have destroyed a test that can give the proof.

I stick around to meet protesters returning from the field. Many won't talk, but others will, and I am surprised when some of them remove their hoods and masks. Two of the people I speak with are Dorothy Cussens, a government nutritionist, and her husband, Chris, who has a computer business. They live nearby in the village of Berrick Salone; she is sixty-one and he is fifty-nine.

When they heard about the protest, they went in search of white suits. In a hardware store they found the sort used for painting and bought two. She does the talking, and in her words I hear sentiments echoed across Europe, where outraged citizens are confronting multinational companies to stop genetically engineered food while they still can.

"We've never done anything like this before, but we feel that this is an unnecessary technology being imposed on us against our will. Nobody has asked for this; it is being imposed on us. And if you damage the environment, you damage humankind," Dorothy Cussens says.

15

WHAT WENT WRONG?

*The progress of technology is continuous; propaganda must voice this
reality, which is one of man's convictions.*
—Jacques Ellul, *Propaganda*

*Monsanto is the devil incarnate and Bob Shapiro walks around in
horns and tail.*
—Gordon Conway, President, Rockefeller Foundation

GREENPEACE HAD A special guest, sort of, at a European planning conference in October 1999: Monsanto Chairman Robert B. Shapiro, who was
speaking to the gathering by satellite. Shapiro had consented to travel to
London for the event, then changed his mind. In any form, his participation was noteworthy, for much had transpired in Europe over two years
between Monsanto and its chief antagonists. Just as remarkable was the
humility of Shapiro's *mea culpa*s.

Shapiro confessed that his company had "irritated and antagonized people" in its drive to bring genetically engineered crops and food to the
Continent. "Because we thought it was our job to persuade, too often we
have forgotten to listen," he said, speaking from an oversize screen to the
right of the dais.

"I think we have tended to see it as our task to convince people that
this is good technology, useful technology—to convince people, in short,
that we are right and that, by extension, people who have different points
of view are wrong," he said.

"Our confidence in this technology and our enthusiasm for it has, I think, widely been seen—and understandably so—as condescension or, indeed, arrogance . . . We are now publicly committed to dialogue with people and groups who have a stake in this issue. We are listening," he said.

By then, it was late for listening, perhaps two years too late. Europe was lost to genetically modified plantings for the foreseeable future, and European-styled debate over genetic engineering had migrated to South America and Asia and was fomenting in the United States. Shapiro's company was on the verge of losing its independence in a merger. His vision of vertically integrated life-science companies providing food, pharmaceuticals, and the essentials of living was on life support.

Unless catastrophes come calling, neither Greenpeace nor any other critic is likely to end the spread of genetically engineered crops. But in Europe, the new technology was stalled. In February 2001, the European Union voted in favor of new and toughened rules for genetically modified food, a step toward ending the *de facto* moratorium that had prevented the technology from taking root in Europe. But the revised 90/220 directive needed approval from national governments and parliaments, and that would require a minimum of eighteen months and perhaps much longer. EU representatives from France, Italy, Denmark, Greece, Austria, and Luxembourg promised that their countries would block new approvals for planting modified crops or importing them into Europe until strict rules for traceability were in place. How and when American exporters would comply with their demands to document the source of grains remained a mystery. European bureaucrats tried to negotiate their way out of the impasse, but the continent remained vulnerable to tricky political winds that blew up typhoons on each new disclosure about threatened butterflies or contaminated taco shells.

Biotechnology's future depends on Europe for three reasons: it is the world's biggest market; it is a gateway to Africa, India, South America, and other lands once under the European flag; and it adheres to the "precautionary principle" in regulating, inviting comparisons with a very different American system that balances risks of technology with their perceived benefits to society.

Shapiro and his ilk claimed to be baffled at what went wrong with the magic seeds they expected to ease farming, cheer consumers, and feed the hungry. Were the forces arrayed in Europe—the Mad Cow Disease, the antiglobalization sentiments, the culture-vultures picking at food—too powerful to overcome? Was the technological premise faulty? Were the risks too great? Might the campaign to bring modified food to the market have been terribly flawed?

A generation of student business leaders and sociologists will cut their teeth analyzing what went wrong in Europe. In Monsanto's European escapades, proponents as well as adversaries of new technologies will find lessons.

Monsanto's mistakes in Europe appeared grievous—so grievous that the life-science giant went from a prosperous independent company on the verge of holding monopolistic power along the global food chain to a subsidiary reviled around the world. Given the depth of hubris displayed in Europe, I think it's entirely possible that the industry might repeat its mistakes in the United States.

Some would argue that Monsanto could not have anticipated the gathering storm back in the spring of 1996, when the European Union okayed importing the Roundup Ready soybeans being sown commercially in American fields for the first time. The scorching public debate, perhaps even sabotage, might have been inevitable with a technology that will reshape food, farming, and global trade while reaching into ethics and religion.

It's just as possible that the company might have forestalled the explosion or, at least, prevented the debate from careening out of control.

So often, Monsanto's problems sprang from faulty communications. Like many companies, it has the genes for manipulation and secrecy in its corporate being. Even as it accumulated its prodigious science talent, Monsanto was perfecting its skills in manipulating regulatory bodies. In the years leading up to its European travails, Monsanto deftly marched controversial products to market without tripping public alarms. Indeed, in the mid-1980s, the company helped create in the United States a friendly regulatory system that sidestepped both a central authority and congressional oversight.

Monsanto wanted government regulation to reassure consumers nervous about genetic manipulation of their food. In the infancy of the technology, the master had kept the customer in mind. But in the 1990s, strategy shifted from attending to the needs of consumers to driving products to market.

In 1993, Monsanto won approval from the U.S. Food and Drug Administration for its first commercial product from genetic engineering: a bovine growth hormone that induced cows to produce more milk. It was a moderate success, but a critical mistake. Rather than introducing a factor beneficial to consumers as the first fruit of the biotech era—perhaps an engineered porcine hormone that would put leaner bacon on the breakfast table—Monsanto chose a product that would deliver profit but offered nothing tangible to consumers.

In the American dairy country of the states of Wisconsin and Vermont, Monsanto counterattacked pockets of resistance with consultants and lawyers. Campaigning in Washington to win federal approval for the cow hormone, the company endured minimal scrutiny at hearings exiled to a suburban Maryland motel. For Monsanto, the revolution in food production was almost too easy, for it instilled a false sense of security. They believed themselves able to manipulate not just genes, but people and governments.

Like many big companies, Monsanto does much of its best work behind closed doors. But in Europe, its strategists discovered that they weren't as clever as they thought.

A RISK-BENEFIT CALCULATION

At the International Institute for Management Development in Lausanne, Switzerland, Ulrich Steger analyzes corporate blunders and advises companies on how to avoid them. Steger has watched companies march forward with blind faith in the power of their rational decision-making. He has seen companies ignore public opinion even when it slaps them in the face. He has seen companies fail because they misjudged emotions that drive consumer preferences.

In Monsanto, he saw all these shortcomings at once.

Steger, who is an economist by training, says Monsanto made fundamental errors. First, the company underestimated the difference between Europe and the United States in attitudes toward food. Speaking by phone from his home in Germany, Steger explained that Monsanto had not grasped how Europeans view the American diet. "In the perception of the European, America is associated with junk food. Even if people like it, it is not the incarnation of culture and sophistication. We're dealing with perceptions and clichés here, but they are important," he said.

Second, he says, Monsanto failed to understand that consumers must feel that they are receiving a tangible benefit if they are to accept risk, in this case the risk of an unproved technology. The first wave of biotech crops were bred overwhelmingly for herbicide tolerance, which would benefit the companies by enabling them to sell more of their proprietary chemicals, and benefit farmers by giving them ease in production, clean fields, and some cost reductions. A second selected characteristic is insect resistance, which leads to fewer chemical sprayings. That benefit might appeal to environmentalists as well as farmers, but not to the masses. There was nothing in the deal for consumers. In Europe, however, consumers found much to be risked, in harm to both the environment and to the well-being of their families.

"If you are a bungee-jumper or a whitewater rafter," Steger said, "you take a risk for a benefit that is visible and important to you. But in all cases where someone else imposes a risk on you, and you don't have a benefit, you oppose it. Why the hell would someone put themselves at risk when there is no benefit?"

Third, Europeans were outraged by the American arrogance and threats. If they didn't capitulate, the United States threatened to bludgeon them with rulings in the World Trade Organization, the global adjudicative office for trade disputes. Consumers rose up in resistance at being forced to take technology they don't trust to produce food they don't need. Europe bridled in the late 1990s as the United States government demanded in the World Trade Organization that Europe accept American hormone-raised beef. Now, the Americans were back with more manipulated foodstuff of dubious value.

"It absolutely puts it in the context of everyone saying, 'oh yes, the Americans once again want to sell us something that is dangerous,' " Steger said.

THE "WINKELRIED EFFECT"

It would further insult Europeans that an American company sought to fly in beneath the radar screen to deliver its revolutionary cargo.

Monsanto did little to prepare the European public for the arrival of a technology that would transform their food and their business. In Brussels in 1998, EuroCommerce lawyer Maria Fernanda told me that she had been stunned by Monsanto's inability to grasp public sentiment for food labeling. "I just hope they change their policies and understand that consumers in Europe should have a say," Fernanda, who represented retailers and importers alike, said.

Retailers in Britain—still smarting from losses when the Mad Cow epidemic made beef unsaleable—resented Monsanto for ignoring allegations that they were sneaking genetically modified soybeans into the European food chain. The brewing revolt split retailers: The British Retail Consortium wanted a boycott of soybeans, while the Food and Drink Federation, which represented manufacturers, did not.

I asked Gene Grabowski, a vice president of the Grocery Manufacturers of America trade association, how, from the food industry perspective, calamity could have been avoided in Europe. "At the very beginning," he told me, "the biotech industry should have sat down with the food industry, and said, 'not everything about this technology is completely understood. You should know this.' And then they should have put five or ten million dollars in ads that say 'this is a new age that is arriving here in Europe with a technology that not only will bring you food but will provide vaccines to your children.' An opportunity was missed. They didn't look far enough ahead."

Monsanto missed the mark again in failing to woo European environmental groups with the chemical-cutting potential of its engineered crops.

They didn't recognize the technology's anti-insecticide promise as an ace in their hand even as it was playing in the United States, placating middle-of-the-road advocacy groups.

In internal meetings during that period, a Monsanto mantra was "freedom to do business." Company officials would say, "They can't stop it," assuring themselves that their victory was inevitable—as it might have been, had the company been able to get modified corn into Europe during the planting seasons of 1998.

Toby Moffett, who was Monsanto's vice president for international affairs until 1999, insisted that his company had no choice but to push hard in Europe after the modified crops began sprouting commercially in the United States. "Once the ball was on the tee, we had to hit it," he said.

But when we met for breakfast at the Willard Hotel in Washington, Moffett made it clear that he knew that his free-swinging company had behaved like hackers sneaking onto St. Andrews Old Course in Scotland. "Monsanto barged in like barging in on someone's private party. We were not European enough. We weren't sensitive enough to European tastes. So we learned the hard way," he said.

As Shapiro acknowledged to Greenpeace, Monsanto's hard-driving ways alienated Europeans. Even British officials who remained publicly in tune with the Blair government's support of the new technology were irked by Monsanto's methods. When I met with British Food Safety Minister Jeff Rooker in July of 1999, he went so far as to blame Monsanto for getting in the way of British progress.

"We invented everything, but we produce nothing here," Rooker told me in his office. "Here we are with a branch of science, biotechnology, on which all our modern medicines are based, and we've got a chance for once in our life to have a science-based economy. Then along comes a multinational corporation with a terrible track record to upset it. The idea that Monsanto, the company that gave the world Agent Orange to defoliate the forests in Vietnam, would set itself up as a life-sciences corporation and tell the world that we are going to feed the world safer food . . ." Rooker's voice trailed off in indignation.

"I only met them once," he said. "They came into this office, almost weeping. I just think they got it all wrong."

Monsanto rivals were irritated, too, I learned at a visit to Novartis headquarters in Basel, Switzerland. The Swiss life-science company's communications director, Arthur Einsele, grew animated on the subject of Monsanto style. "They come into a room with their ten people and their machine and push, push, push . . . They cause problems with their attitude," he said.

As the global leader in food biotechnology, Monsanto had the most to gain. It also had the most to lose, which may explain why European companies a year or two behind the American leader refused to leap when it said jump. When Monsanto tried to recruit its European brethren into advertising campaigns, the European companies hid. Monsanto remained strikingly alone, the global face of biotechnology and the object of public ridicule from Europeans who seemed to forget that some of the world's leading biotech firms are based in Europe.

I recall a cartoon in a British magazine, with one character saying to another: "You say tomato, I say Monsanto."

Blundering forward, oblivious to foes and resentment, Monsanto sealed its fate. Europeans have a word for such strategies. They're called the Winkelried effect, after Arnold Winkelried, the legendary Swiss hero who blocked Roman invaders by stationing himself in the middle of the road. "All spears on me," he shouted. In short order, Winkelried lay dead, perforated by all those lances.

Then the trailing forces counterattacked the Romans, who were now short on spears. As the poem goes:

> *"Make way for Liberty," he cried.*
> *Made way for Liberty and died.*

FINALLY, ADVERTISING

In June of 1998, Monsanto unleashed its version of spears: solo ad campaigns that became the talk of the continent. By the standards of American attack ads, these were tame. But they were extraordinary in Europe, where aggressive ads aimed at fickle consumers are considered risky.

For three months, in the Sunday supplements of British broadsheets and in leading French magazines, the ads trumpeted the grandest hopes of biotechnology. They wisely emphasized the capacity of genetic engineering to slash the use of pesticides. In one of the British ads, displaying a pastoral scene of crops beneath blue skies, Monsanto asserted that "we believe plant biotechnology can limit industrial and chemical impact to the earth."

The ad was bold, coming from a company selling chemicals for ninety-seven years. In both countries, the advertising endorsed labeling modified foods. In French magazines, alongside a photo of a cantaloupe, an ad declared: "You have the right to know what you eat, especially when it's better . . . After several months of debate, Europe has just adopted a new law for the labeling of food that comes from genetically engineered plants . . . We believe that products that come from biotechnology are better and that they should be labeled."

Those assertions might have confused Americans, knowing Monsanto as a company opposed to labeling modified products in the United States. But as people had told me across France—where opposition to GMOs did not, in my view, run deep—their support for modified foods hinged on knowing their ingredients.

Yet Monsanto was widely scorned for its initiative. A failed ad campaign became part of the accepted wisdom on why Monsanto lost Europe. As late as June 2000, the American magazine *Business Week* attributed what went wrong to Monsanto's having "mounted an ill-conceived public-relations campaign in Europe to head off growing criticism of genetically modified seeds. But the effort seemed only to throw fuel on the fire."

Indeed, Monsanto's brief campaign was viewed as such an unmitigated failure that for years its specter constrained advertising throughout the biotech industry.

The ads may have flopped, but not for the reasons usually cited.

Monsanto was plagued by bad timing, no doubt about it; the day the company announced its campaign was the day that Prince Charles delivered his withering denunciation of genetic engineers for playing God. By then, Monsanto was already late. Faulty instincts and an opposition faster on the draw had already shaped popular perception. The biggest mistake of all, as I see it, was the company's decision, in the autumn of 1998, not to renew its ad campaign.

I am convinced that had the ads come sooner and stayed longer, they would have worked.

In the early days of a new technology, the Monsanto ads aimed to fill a critical void: persuading people that altered foods contain benefits. In France, for example, the ads asserted that modified tomatoes could be kept on the vine longer and therefore allowed to ripen and acquire their full taste rather than be picked early for shipping. Alongside a plump strawberry in a British Sunday supplement ad, the text read: "Funny how we all suspect fruit and vegetables don't taste as they used to. Year-round demand, forced ripening times and early harvesting are to blame. Plant biotechnology offers the potential to produce crops that not only taste better, but are healthier."

Monsanto also delivered its controversial feed-the-hungry appeal: "Worrying about starving generations won't feed them. Biotechnology will," a British ad began.

Accompanying Monsanto's ads were new Internet sites that not only included unfavorable news accounts about the company but also linked to web sites of the company's detractors. Even Greenpeace. It was a new, frank approach reflecting the hope that Europeans would react favorably if given more information, even critical information. So appealing was the message that France's socialist government recommended that schools use the Monsanto web page.

"Why not just step up?" I quoted Philip Angell, then Monsanto's director of corporate communications, in the *St. Louis Post-Dispatch*. "If we don't have confidence enough in the facts about biotechnology that they can stand up to other things being said about it, then we shouldn't be in this business." Unfortunately for Monsanto—and fortunately for its critics—the savvy Angell did not stay with the company.

The company devoted just $2 million to its twelve-week campaign, a fraction of what is typically spent to launch new products. In Britain, pollster Stan Greenberg concluded in his famous leaked analysis that the ads were "for the most part overwhelmed by the society-wide collapse of support for genetic engineering in foods. The advertising did not penetrate a large portion of the [intended audience] and, among those who read the advertising, it did little to increase public acceptance of GMO foods." But Greenberg had detected no evidence that the ads had backfired; on the contrary, he concluded that they had improved the company's standing with some important segments of the public, buttressed support in the government and parliament, and showed retailers that Monsanto was trying to win consumer confidence.

In France, the evidence was even clearer. Greenberg's polling, never revealed at the time, showed that the ads clearly had worked. In the fall of 1998, support was deteriorating in France, as in the rest of Europe, Greenberg concluded. But the ads helped to contain the slide; they were keeping Monsanto from becoming an issue itself in France, as it had become in Britain. And that was a victory in a country that maintained a healthy respect for science and that, as Europe's breadbasket, held the key to the entire continent. In other words, Monsanto and biotechnology still had a fighting chance in France.

"The advertising campaign has made significant penetration and has significantly increased public acceptance" with a key segment of French citizens, Greenberg wrote at the time. "It has given greater credibility to key arguments in favor of this new technology . . . Monsanto's French advertising campaign was an immense success."

But the ads never ran again in France. Nor were the British ones revived. Company officials in St. Louis remained averse to trumpeting biotechnology's capacity to reduce chemicals. Perhaps they had concluded, after unremitting accounts of the European revolt, that their cause was already lost. Maybe they never understood that a political battle needs political weapons. Whatever the cause, communication was the failure.

In October 1998, during a stormy meeting at Monsanto's European headquarters in Brussels, visiting company officials vehemently refused recommendations to renew advertising. "They ground us into the floor of the Monsanto office," said one company official who was present. The response of one of the St. Louisans suggested that Monsanto was about to climb into a bunker.

"Are you fucking nuts?" said the St. Louis official on being presented with a menu of new advertising options.

THE POLLTAKER

On the walls of his Washington office behind Union Station, Stanley Greenberg displays his connections. There are campaign posters of U.S. Congresswoman Rosa De Lauro, D-Conn., who is his wife. There is a huge drawing of Bill Clinton, for whom Greenberg handled polling before Clinton's first election as president of the United States in 1992 until the spring of 1995. In a conference room with a chandelier resembling inverted champagne flutes, there's a poster of Ehud Barak, who took counsel from Greenberg on his way to becoming Israel's prime minister.

Greenberg has worked for clients in, among other places, France, Germany, Brazil, and Argentina. For a long time, he spent at least one week every month in Britain, taking polls for Prime Minister Tony Blair, the Labor Party, and corporate clients. He polled for Vice President Al Gore in Gore's failed presidential campaign. He has taken surveys around the world on radiation and a host of public policy matters. But in his many travels, never has Greenberg seen an issue roil the waters like genetically modified food.

In Britain, Mad Cow Disease had ignited flames of fear, and trust in institutions had gone up in the smoke as a pyromaniac media fanned the fire. Still, Greenberg said, "The scale of this has been difficult for me."

Greenberg sees miscalculations from the beginning by his client, Monsanto. And given what he was still seeing in the new century, Greenberg was concluding that the biotech industry had not learned its lessons.

Just a few days before we met, in June of 2000, a *New York Times* story had revisited the European backlash against GMOs, focusing on farmers' worries about pesticides. From his client's perspective, Greenberg spotted something missing. "All it required was somebody from the industry asking the farmers, 'why do you prefer growing potatoes that, almost more than any other crop, have been dunked with chemicals? We'd rather give you potatoes that are free of chemicals.'"

Greenberg compared his findings on opposite sides of the Atlantic. "There's an innocence in this," he said. "It's an American innocence that's part of Monsanto, and it's an endearing innocence, which is the idea that global problems such as sustainability and hunger can be addressed through abundance, through increased production. That's a very American notion, and in polling in the U.S., it scores very well.

"One of the reasons for this technology is to help parts of the world that are not well endowed for agriculture to produce more food to help their people. But Europeans just don't believe that; they don't believe that hunger is a function of lack of sufficient production. They believe it has to do with politics, corruption, and distribution. There's a very different view of the world in Europe and in the United States."

I knew the brainy, diminutive Greenberg from having covered presidential campaigns. But I was unprepared for his candor, especially because, on occasion, he still worked for Monsanto. He asserted that Monsanto had opened in Europe with a faulty game plan from which it never recovered. The European plan, like Monsanto's American strategy, was based entirely on farmer acceptance: persuading farmers that using the bioengineered seeds would mean less tillage and fewer chemicals.

"The assumption was that they didn't need to win consumer confidence in this process and that, ultimately, the products would just make their way through, that farmers would adopt them on a large scale and there wouldn't be any public concerns, any regulatory concerns. Demand was left entirely out of the process. That was also true in the U.S. The problem is that we're operating in a global world, and there are global consumers and global regulatory agencies. Consumers were never in their strategy. To me, that was the most fundamental problem.

"The second problem was that the company never developed a self-confident vision of the benefits. They were constantly immobilized by an inability to say that 'this product reduces pesticide use.' This company made a transformation from being a chemical company to a biotechnology company, and they did it very impressively. However, they continued to sell chemicals, and they were internally fractured on being able to articulate the benefits.

"The most important thing for them was to have come down on the side of the benefits from their products and argue with confidence. You can't expect the public and press to stand there in support of your technology if you don't say with confidence what the benefit of it is."

You don't need to be a genetic engineer to understand Greenberg's summation: "Monsanto is just awful at making their case. And that's part of this story: the arrogance of science."

CURIOUS STRATEGIES

I was oddly relieved after Greenberg confirmed my suspicions about the roots of Monsanto's European collapse. I felt a tad vindicated after having been scolded by some in the company for writing critically from Washington, Europe, and elsewhere in the world about its biotech failings. The company had not appreciated accounts of their worldwide travails on page one of the hometown paper.

My reporting had made my relationship with the company touchy. I kept channels open with scientists, and some officials at the company would return my e-mails, answer my phone calls, and accept my visits. I called the public-relations people when I needed "comment." But with a handful of exceptions, the company sent me no press releases and initiated no contact with me for over two years. This included the summer of 1999, a superheated period for biotechnology news, when I was routinely sought out by foreign reporters and broadcasters and gave presentations on biotechnology to visiting delegations of opinion leaders.

Robert Shapiro, Monsanto's chief executive officer and the modern architect of genetically modified foods, never would sit with me for an interview.

I've never devoted much time to analyzing relationships with people I cover. I begin from the premise that politicians and companies have the right to define their own press operations, and that journalists must play by their rules. At the same time, I've noted that skilled publicists with the latitude to cultivate relationships and deal openly with journalists are the most successful. That's been my experience in Washington in public policy and in presidential campaigns back to Ronald Reagan's re-election in 1984.

Typically, companies and campaign organizations reflect their leader. And leaders who understand how the news business works—the exigencies of deadline and the fundamental truth that reporters report what they have in hand—can make the media work to their advantage. For whatever reason, these basics were Greek to Shapiro-led Monsanto. I have no doubt that attitudes toward the press accelerated both the company's slide and Europe's rejection of modified food.

Monsanto is by no means alone among companies in distrusting the news media. (A friend in the corporate world told me that prying reporters are simply hated. "There's no other way to put it," he said.) Monsanto exhibited its disdain for reporters regularly, either by failing to cooperate or taking so much time to do so that it would be too late to shape the story. For instance, Shapiro broke several engagements to meet with the *New York Times* editorial board in 1998 and 1999. Other chief executives might have understood that stiffing the *Times*, which often sets the tone for American news coverage, is perilous.

In 1999, Shapiro's tendency to disappear from the fray was striking. Many in the company, including scientists, wondered why he was not more involved. Some speculated that his recalcitrance had its roots in the humiliating experience of being hit in the face with a pie while speaking in California in October 1988.

A company needs a chief executive out front defending the product, answering the skeptics. If the president isn't available, someone else must wear its public face, taking chances in forums and speaking with people through their emissaries, the news media. At critical times in the late 1990s, Monsanto had neither. "He [Shapiro] was AWOL," said one former company official. "As a result, we didn't have the guts; we didn't have the courage of our convictions."

The heads of Monsanto and Microsoft Corporation, two corporations in the news during this period, reflected opposite approaches. Both companies were leaders in bringing important technologies to the world. Each endured searing heat at the same moment in history: Monsanto from activists and consumers who, unfairly I thought, came to regard the company as the embodiment of evil; Microsoft from the United States government for alleged monopolistic business behavior. Shapiro shrank from public engagement, while Microsoft's Bill Gates seemed to the world omnipresent, irrepressibly defending his company.

Shapiro wouldn't talk to me, but he spoke on occasion in 1998 and 1999 to Michael Specter of the *New Yorker.* I thought some of his remarks were telling. "There are two things that most of us feel. We feel hurt and we feel angry," he told his interviewer. "We had real leadership; we had worked hard to do it. We had shown faith in this science when others were dubious, and it all seemed to be working. So we painted a big bull's-eye on our chest, and we went over the top of the hill."

In something else he said, I think Shapiro got it right. "The reason there is controversy about this has nothing to do with biotechnology," he said. "This is about power. It's about them saying that if you want to make changes in people's lives or introduce new technology, you are going to have go through us. And if we don't approve, we are going to bring you down."

In an essay Shapiro wrote in early 2000 for Washington University's Center for the Study of American Business CEO Series, the soon-to-be-retired Monsanto chief said he learned that "there is often a very fine line between scientific excitement and confidence on the one hand, and corporate arrogance on the other. It was natural for us to see this as a scientific issue to be decided by scientific experts. We didn't listen very well to people who insisted there were relevant ethical, religious, cultural, social, and economic issues as well. There is a huge difference between food as science and food as culture. Food occupies an important place in many cultures and countries well beyond the necessity of sustenance, and carries an almost inexpressible emotional resonance."

He ended his essay, humbly I thought, by quoting a past president of the National Academy of Sciences. Philip Handler's words, it occurred to me, probably wouldn't have resonated with Shapiro a few short years ago, as he and his lieutenants tried to kick in Europe's door. "The so-called technological imperative—that which can be done, will be done—is not an inviolable natural law. It is indeed for the larger society to determine whether or not whatever it is will be done."

Turned down by him for interviews, I tracked down Shapiro's e-mail and gave that a try. In my message, I listed a few of the blunders for which his company is criticized. As an afterthought, I noted that one of his allies regarded him "as a good man traveling on bad advice."

That may have been the phrase that prompted a reply.

Shapiro responded that he had made most of his company's key decisions and, given the facts and the alternatives at the time, he believed that his actions made sense.

He said in the e-mail that things always look different in hindsight, but that he would stand by what Monsanto had done. Ultimately, he wrote, the test would be the answers to three questions: whether his company had conducted its dealings ethically and honorably; whether biotech would, indeed, revolutionize global farming; and whether Monsanto would remain a leader. He was pretty sure that the answers to all those questions would be yes.

I had asked Shapiro how he might be viewed when others look back on the turbulent times when genetically modified food was introduced to the world. He replied that aside from the people who knew him well and whom he cared deeply about, he doesn't much care how he is remembered.

MONSANTO'S NEW LEAF

A few months later, in November of 2000, Shapiro was gone from what was left of Monsanto and the company's new president and chief executive officer, Henry Verfaillie, stood behind a microphone at a Washington, D.C., hotel trying to make amends. Saying that his company was "blinded by enthusiasm," Verfaillie promised a new era of open dealings and of dialogue with skeptics of biotechnology.

"We missed the fact that this technology raises major issues for people—issues of ethics, of choice, of trust, even of democracy and globalization," Verfaillie said in his speech to a *Farm Journal* conference.

"When we tried to explain the benefits, the science, and the safety, we did not understand that our tone, our very approach was seen as arrogant," Monsanto's Belgian-born leader said.

By now, there was less to be arrogant about. A month before, Pharmacia had spun off Monsanto, and the newly constituted company had issued its initial stock offering. For the once-mammoth life-science company, it was the opportunity for a new beginning.

"Because we are a new company, we have the opportunity to change—to change our behavior and our actions—and to be measured on how well we do it. We have decided that the 'something' that must change is us. And so we are knowingly and deliberately choosing a new path," he said.

Verfaillie said his company had done many things right in introducing its technologies. He added, "I would like to be able to say that this story ends with 'and they lived happily ever after.' But that, of course, is not what happened. The company—my company, Monsanto—had focused so much attention on getting the technology right for our customer, the grower, that we didn't fully take into account the issues and concerns it raised for other people. We thought we were doing some great things. A lot of other people thought we were making some mistakes. We were blinded by our own enthusiasm."

He gave his words a name: the New Monsanto Pledge. Ten years before, the company had declared its initial Monsanto Pledge, a commitment to environmental responsibility.

Given Shapiro's similar words in that meeting with Greenpeace a year earlier—he, too, had used the word "arrogant" to describe his company's misadventures—I wondered about the latest *mea culpa*s as they echoed through the speakers of an underground ballroom. Afterward, I reminded Verfaillie that Shapiro, too, had said these things. What was different, I asked.

Verfaillie, dressed in a navy, double-breasted suit, looked at me through a knot of journalists. "Me," he said, raising expectations even further for the company that had generated a global furor. "What is different," he said, "is that we have said that not only are we going to talk about these things, we are going to do something about them."

PLANTINGS EIGHT 〜

In my experiment with bioengineered soybeans, I am not fooling around. Today I sprayed putrefied egg whites to drive away deer that might devour my plants. The four-legged vandals from the woods across the road have already stripped young tomato and eggplants. If the smelly concoction doesn't work, I've got a bottle of one hundred percent coyote urine in the shed.

I also spread nitrogen fertilizer, which worries me more than animal pee. In our century's passion for bigger yields and evergreen lawns, we have nearly doubled the amount of nitrogen on the planet. All this nitrogen creates algal blooms that destroy oxygen and choke aquatic life—from Japanese waters to the Adriatic Sea. Once I steamed with scientists into the heart of a seven-thousand-square-mile "Dead Zone" in the Gulf of Mexico to measure the effects of the nitrogen that runs off Midwestern farms and down the Mississippi River. Nothing lived in this massive blob of shifting, hypoxic water, and at the bottom, which was layered in white bacteria, even the worms lay dead.

Genetic engineering has been touted as a cure for this problem by enabling plants to achieve "nitrogen fixation." Soybeans and other legumes such as alfalfa and clover already have some capacity to fix nitrogen, which means they convert atmospheric nitrogen into compounds they need to live and grow. Beginning in the mid-1980s, we were told that genetic engineers would create nitrogen-fixing plants, thereby eliminating the need to inundate our world with nutrient pollution.

That goal remains elusive. "Nobody has a clue how to do it," an industry scientist told me candidly. Take corn. To engineer a nitrogen-fixing corn, scientists would need to transfer about two dozen genes, many of which have not yet been identified.

Some of the genes would help to produce the enzymes to fix the nitrogen. But others would have to be inserted to protect the enzymes from oxygen, which exists with nitrogen in the environment. Once you figure out which genes you need and how to make them work, there's a good chance that corn plants engineered with them would be spending so much of their energy dealing with nitrogen that they would yield less corn.

If your crop takes a yield hit, you've traveled from science into the realm of economics, a determining factor in whether any sort of invention takes hold. With synthetic nitrogen cheap, why would farmers risk growing fewer bushels per acre?

Me? I'll take any risk necessary to see what these genetically engineered soybeans are about.

16

IN CYBERSPACE, TECHNOLOGIES CONVERGE

When spiders unite, they can tie down a lion.
—Ethiopian proverb

IN THE SUMMER of 1999, Charles Benbrook, a consultant in Sandpoint, Idaho, arranged to post on the Internet the Deutsche Bank's bleak analysis of the future of life-science companies. He e-mailed the web address to fifty people. The report, written during the depths of the technology's European collapse, concluded that the future of genetically modified foods was, at that moment, grim. It didn't matter, the bank analysts observed, that the technology appeared safe or might hold promise for the environment.

"The perception wars are being lost by industry," the analysts wrote. "Consumers may very well decide that biotechnology derived foods are not as appealing as organic or current offerings."

In three days, thousands of people had retrieved the twenty-five-page report and disseminated it around the world. One year afterward, over one hundred people weekly still were downloading it. Investors, farmers, critics, and citizens making up their minds about a confusing technology had a fresh and damning piece of information.

As recently as the mid-1990s, the analysis, which was a service to the bank's clients, would have been seen by few people. Even if it had escaped the circle of investors, it wouldn't have traveled far. But that was before the Internet, which has enabled information to be spread and people to be mobilized at nearly the speed of thought.

In the early going, the Internet gave biotechnology companies fits.

Genetically modified foods stalled on the way to the global marketplace for many reasons. In Europe, gateway to the world, Monsanto planted their wonder crops before cultivating trust. Not only Monsanto but the entire life-science industry failed to foresee people's natural concern about a powerful technology, especially a technology developed to benefit the interests behind it—who happen to be the same interests testing its safety.

There was another reason, a force often overlooked in biotechnology's late 1990's tumble: the Internet. The debate over genetically modified foods began in earnest just as the number of people going on-line around the world skyrocketed. In 1996, the first year of modified plantings in the United States, Internet connections globally expanded by 50 percent. By 1998, as transgenic crops soared to seventy-three million acres, the number of Internet hookups had again doubled.

Two transforming technologies emerged at the same historic instant. But in their infancies, one stunted the other. The new global marketplace of ideas proved more potent.

By breaking the monopoly of experts and publishers, the Internet revolutionized how we communicate. After watching the biotechnology debate up close, I believe that it will be recalled as the first global public policy war of consequence to be waged on the World Wide Web.

In the past, people fearful about technology revolutions had no ready, affordable means to multiply their voices. Often, their audience was small, their medium slow, and their approach amateur. So when their misgivings were heard, the worriers were branded as Luddites, radicals, or eccentrics opposed to progress. As biotechnology emerged in the 1990s, the Internet had evened the odds.

Chuck Benbrook, the cyberspace *provocateur* in the rural Idaho enclave, hasn't always lived so remotely. He worked in the White House during the Carter administration, then on Capitol Hill, and later as the executive director of the board of agriculture at the National Academy of Sciences. For eighteen years, he was a player on farm issues in Washington. But thanks to the Internet, Benbrook could affect a global debate every bit as forcefully from the Idaho Panhandle as from the science academy's Foggy Bottom offices. All it took was a bit of chutzpah—talking the Deutsche Bank out of its report—and a few minutes at the keyboard.

When I asked him about the Internet and biotechnology, Benbrook recalled the struggles over pesticides in the 1980s, when the chemical industry was swamping farmers, government agencies, and research institutions with prochemical information in order to water down legislation in Congress. Advocates of biological controls and less chemically dependent solutions, Benbrook said, "didn't have a way to combine their voices and knowledge and experience. As a result, the whole scientific effort was marginalized for a decade and a half, and it wasn't until the mid-1990s that there began to emerge a viable alternative to sole reliance on chemicals."

The Internet, Benbrook asserted, "is to biotech what genomics is to breeding: It has vastly expanded the sources of information. It makes it possible to bring major categories of important information and literature to the world community, something that could never have happened before. Now the activist community following biotech knows a lot more about what is going on and what concerns people might have.

"Activists can transfer fresh and important information around the world with speed and ease. And that's something we've never experienced before," said Benbrook, who has become an activist himself since his Washington years.

Benbrook said something else that applies to both sides in the GM food wars: "If the public doesn't believe what is said, the fanciest web sites and the biggest public-relations campaigns in the world won't amount to much."

TRACKING THE ENEMY

Before 1998, eight thousand was the most public responses the Department of Agriculture had received on any subject. Then-agriculture secretary Dan Glickman asked people to tell him what they thought of a proposed organic-foods policy that would allow products to carry the government's new organic certification even if they'd been genetically engineered, irradiated, or fertilized with sludge.

Over 275,000 people responded, nearly all of them by e-mail, the vast majority outraged at the government's effrontery. Monsanto, too, was stunned, and in a letter to Glickman, the company backed off making a public fight out of its demand that modified foods fit into organic labels. As a result of the e-mail uprising, Glickman declared in May of that year that food labeled organic could not originate in the labs of the genetic engineers.

What I was seeing, I suspected, was not just a new way of organizing but a new way of governing. I asked Phil Noble, a political consultant and founder of PoliticsOnline, what he thought. "It changes the presumptions of representative democracy," he said, speaking of the Internet as a political tool. "I think the Internet is going to do for public policy what the telephone did for lobbying."

Noble added a warning about the danger of manipulation in e-mail campaigns. "In literally a matter of hours," he said, "I can create an interest group of tens of thousands on whatever my issue is right now, and mobilize them to send mail, e-mail, or even rotten eggs."

The issue mobilizing environmental and consumer activists in the mid-1990s was genetically modified food, and the Internet brought them together, empowering communications on any budget. With their e-mail list-serves, they distributed every damaging news story and morsel of information raising questions about the technology. I receive at least a dozen stories on my e-mail every day; a recent study pointing to diminished yields from modified crops was sent to me six times. (Later, in 2000, a speech by Microsoft's Bill Gates praising biotech was sent to me five times by the technology's supporters.)

The Internet has done little to help activists gain the access to the political elite that is enjoyed by companies who pay for that privilege. Yet it is having a massive effect in policy sanctums and company boardrooms.

With a paid staff of five in the United States and Canada, the Rural Advancement Foundation International, better known as RAFI, had about thirty thousand fewer employees than Monsanto. But RAFI's Internet campaign against the Terminator created a monumental public-relations headache for Monsanto and fueled anti-biotech feelings around the world. RAFI even coined the name, Terminator, by which the so-called Technology Protection System has come to be known. As patented by the Department of Agriculture and a Mississippi seed company, the technology renders seeds sterile, so they can't be saved for the next crop.

In a sixteen-month period, RAFI had 1.3 million hits on its web site, from which people downloaded 450,000 pages—many of them about the Terminator. "The Terminator campaign never would have been possible without the spread of information on the Internet," RAFI's Hope Shand told me. It was so successful that middle-of-the-road research institutions and even the biotech-friendly Rockefeller Foundation condemned the technology on the grounds of threats to low-income farmers. (Monsanto later disavowed the Terminator.)

I have a telling chart, compiled inside the biotech industry. It contains the web addresses of more than one hundred groups regarded as critics, the names of their members, details about the site registration, the groups' histories, and who links to whom. The industry devotes high-priced talent to monitoring anti-GMO activities, partly by using e-mail addresses that do not divulge their companies' names. In the new century, industry strategists believed that they were monitoring enough sites to know when studies critical of their technology were about to surface.

"Now, we're able to pick these up well in advance and get our scientific evidence together so, unlike the monarch butterfly, they never get any traction," an industry insider boasted.

After getting zapped in cyberspace, the biotechnology industry concluded that the time had come to know the enemy. Web addresses on the chart include those of middle-of-the-road organizations such as the Union of Concerned Scientists, which asserts on its web site that genetic engineering can play a "minor but useful" role in delivering new products. Organizations on the industry chart are as large as the Green Party in Britain and the Sierra Club and as small as Benbrook's one-man band in Idaho.

Special attention goes to militant groups. When I type in some of the web addresses, I am propelled on a journey through a global circus of anti-GMO activity. There's the Ruckus Society, the California-based organization that sponsors training sessions in civil disobedience; Genetix Snowball, which promotes destruction in fields of modified crops; and an Internet site called Watching Monsanto, which asserts that "within ten or twenty years, they may inadvertently and unintentionally destroy civilization as we know it, with massive famine and/or world war."

Then there's Mutanto, a web site that parodies Monsanto's. Instead of

Monsanto's slogan of "Food, Health and Hope," Mutanto offers "Fraud, Stealth and Hype."

I have never figured out who, outside of the combatants, has the time to peruse these web sites or exactly how they shape people's thinking. The biotech industry thinks their critics have succeeded through distortion: distorting the facts about safety and creating the false impression that consumers, not just activists, worry about modified food. Company officials believe that relatively few activists have been able to create a sense of movement that wouldn't exist without the web.

"It's a dual-edged sword," Monsanto's Jay Byrne, one of his company's more politically savvy operators, told me. "On the one hand, the Internet allows people with opinions or even spurious facts to share that information broadly. But at the same time, it allows the public access to scientific and academic information that so far has been generally supportive of the technology. The challenge lies in discerning between the two."

THE INDUSTRY ARMED

It took a while, but by 2000 the industry had begun catching up in tactical deployment of the Internet. Monsanto's web sites around the world foster debates about the technology; sometimes they even go easy on the propaganda or omit it altogether. For journalists, the Monsanto United Kingdom site contained a wealth of undiluted news accounts of the rebellion in Britain. On the Monsanto India site, readers can understand how that country is torn between self-reliance and seeking solutions for its people. In France, the socialist government recommended that schools use the Monsanto site to teach students about biotechnology.

A web site orchestrated by the Grocery Manufacturers of America and the food industry opened with this come-hither message: "Imagine your child eating a banana enhanced with vital oral vaccines instead of receiving a shot, fruits and vegetables that contain more beta-carotene and vitamins C and E, and a new grain under development to counter vitamin A deficiency, the leading cause of blindness in children in the developing world. These and many more are the promises of biotechnology."

Pro-GMO warrior Gene Grabowski, vice president of the grocers' group, observed that fifty or so communications every day from scientists and allies show up on his e-mail. "You can share information and respond quickly if someone puts out misinformation. It's so easy. Think of what Mohammed or Jesus could have done with the Internet," Grabowski mused. A web offering by the Council on Biotechnology Information— the life-science companies' joint collaboration—sums up the case that the industry is seeking desperately to make: "Biotechnology is a safe way to produce healthier food in greater quantities, have a cleaner environment and aid in the fight against world hunger."

Well said, but in this era of mounting distrust of corporations, I can't see how the companies' elaborate sites move people who haven't made up their minds. Nor, because they sermonize to the choir, do I view them as an effective organizing tool.

That's not to diminish all the probiotech Internet offensives, particularly the unlikely effort operated at Tuskegee University in Alabama by an India-born crop breeder who can be critical of the companies he's helping.

With his AgBioWorld.org, Channapatna S. Prakash had, by the summer of 2000, organized over twenty-five hundred scientists from around the world for his "Declaration of Scientists in Support of Agricultural Biotechnology." Among those who have taken a place on the Prakash bandwagon are James Watson, recipient of the Nobel prize for discovering the double helix of DNA, and Norman Borlaug, another Nobel laureate who wears the title "Father of the Green Revolution" for boosting food production in developing countries, among them Prakash's homeland.

Prakash runs a chat room, too, where scientists get beyond moaning about their vilification to try to understand the complicated social questions underlying genetic sciences. Prakash's efforts have made him an oft-quoted celebrity of the biotech wars and won him invitations to carry the industry's banner at biotech forums around the world. In June of 2000, for instance, Prakash warned in London that Britain was becoming less self-sufficient in food production, which he attributed partly to its robust opposition to gene-altered crops.

Prakash's Internet labors were, as best I could tell, the most important web effort in the biotech industry. More than anyone else, he has melded together the segment of society missing in the early GMO wars—the scientists, who, out of ignorance, arrogance, or aversion to public discourse, kept quiet.

In a phone chat, Prakash talked of his methods and motivations, while telling me a bit about himself. He was born in 1957 in the southern India city of Bangalore, the south Asia high-tech haven where Monsanto's center of research is situated. He grew up poor in what he refers to as a "port-to-mouth existence." In other words, he ate the food that arrived in Bombay in the form of aid from industrialized countries. The Green Revolution has helped his country immensely, he argues, quadrupling grain production since the 1960s. He believes genetic engineering is the next necessary step. "This is a technology that can help a farmer growing millet on one acre in India as much as it can help an Iowa farmer with five thousand acres," he says.

"Our failure to explain things to the general public got us into the position of people thinking we are some sort of Frankensteinian monsters trying to foist these inedible and dangerous foods down the throats of the public," Prakash continued. "My goal in starting this whole effort was to catalyze the academic community and convince them that we have a responsibility beyond our laboratories."

His followers do more now than just talk to one another. He explained what would happen if, for instance, a "nasty story about biotech" shows up in the *St. Louis Post-Dispatch* under my by-line. "I will call up my friends in St. Louis, not at Monsanto, the big guy in town, but maybe at Washington University, and they will write letters to your newspaper," he warned.

The Internet, Prakash concluded, empowered his opposition. Now it is enabling his own crusade. "There's no way there would be all these protesters without it. The Internet is next only to the invention of language in its power. I think historians in the future are going to say that the Internet was awesome in democratizing communication and access to information. That's the beauty of the Internet. But you can use it for good or bad," he said.

17

IN INDIA, A FATAL CONNECTION

*We have not inherited the world from our forefathers. We have
borrowed it from our children.*
—Kashmiri proverb

*Remove the villager's chronic poverty and illiteracy, and you will find
the finest specimen of what a cultured, cultivated free citizen should be.*
—Mahatma Gandhi

ON THE ROAD to Warangal, the land of dying farmers, I join a traffic jam:
boxy Hindustan Ambassadors; diesel buses and lorries; motorized rickshaws and phalanxes of scooters. Traffic swells and shrinks like amoebas.
Then the flow ceases, thrusting me so closely against fellow travelers in
open conveyances that I can see the hair on their toes.

There are rules in Indian traffic that outsiders don't get. Flashing red
signals are nothing more than gentle encouragement to reduce speed. But
even India's most heedless drivers display an evolved sense of collision
aversion: At the brink, they suppress anger and brake. Then they lurch
forward to the next entanglement.

I have traveled to India to observe the convergence of Western science
with the developing Eastern world. It's not an easy meeting. At one of
its fronts, enemies India and Pakistan are flexing their nuclear might. Both
have exploded test weapons. Another technology war is brewing in India,
from the Himalayas in the north to the flat river valleys of the south. It

is the war over whether Western life-science companies will sow genetically modified seeds in Indian soil.

Genetic technologies, arriving as India prepares to overtake China as the world's most populous country, have triggered a bread-versus-freedom debate. Its resolution may well help answer one of the millennium's urgent questions: Can a chaotic democracy with makeshift ruling coalitions deliver the basics of life to a billion people?

This debate is quickening. The leader of a grass-roots organization claiming ten million farmers is promising an "action," which has many people on edge. When I talked with M.D. Nanjundaswamy, he spoke cryptically about his organization's need to demonstrate its anti-GMO sentiments, and soon. The group, Karnataka Rajya Raitha Sangh, is a force not to be ignored. Two years ago, its members orchestrated the destruction of a Kentucky Fried Chicken in the southern city of Bangalore. Dozens of farmers surrounded the fast-food outlet and smashed its windows. KFC's transgression: multinational intrusion. Before that, Karnataka members occupied the offices of Cargill, the U.S.-based seed giant, in two cities. Now their attention is riveted on Monsanto, the company that is aggressively bringing its genetic technologies to India.

Nanjundaswamy, the group's founder and leader, is upset with the government's decision to permit Monsanto to plant cotton in the first GMO field trials in his country. He considers the technology an imperialist weapon, and he tells me that farmers believe that the Terminator is coming to India's fields. I tell him that I am reasonably certain that the Terminator is years from production. Certainly it is not engineered into the seeds that have sprouted into knee-high bushes in India's test plots, as is the rumor. He says he understands this, but that rumors circulate nonetheless.

Nanjundaswamy, a law professor and an influential man, traces the roots of the organization he founded to the ideology of Mahatma Gandhi, who viewed rural villages as instruments for revolt against British rule, which finally ended in 1947. In Gandhi's *swadeshi,* which translates to local goods only, Indians in search of freedom rejected cloth from the English textile mills in favor of the courser homespun *khadi.* Listening to Nanjundaswamy, I began to doubt that he followed Gandhi's path to nonviolence.

"We will physically throw them out," he said of Monsanto.

DEATH IN THE FIELDS

Heading toward Warangal, I notice placards that read: "To the Greatest Visionary and Missionary." I know what they are about because this morning in the Katakaya Hotel in Hyderabad, before I departed for Warangal, I peeked into a room guarded by a Western security corps. There, eating breakfast, was Pat Robertson, the American evangelical broadcaster. In southern Virginia, where he lives, Robertson is fond of moving about in his black Corvette. In Asia, he is traveling in an airborne "Hospital Ship" dispensing health care along with spiritual feeding. Robertson arrived too late to help the farmers of Andhra Pradesh.

In a little more than a year, over four hundred farmers in this single Indian state have committed suicide. Some hurled themselves down wells hundreds of feet deep. Their relatives had to retrieve their bodies by dangling the equivalent of meat hooks into the blackness. More often, the farmers swallowed cocktails of insecticides, then lay down to die in their fields.

Farmers also are killing themselves in Punjab, India's breadbasket to the north, and in the states of Harrea, Mkaharta, and Kantati. All told, at least seven hundred Indian farmers have died by their own hands in a year's time, I am told reliably, and it may be many more. In a drought a decade ago, two hundred farmers killed themselves. But nothing in recent times matched the season of despair that had swept across India's rural lands. Something new, something unfathomable to me, had taken root.

Rural India is an arena for disaster. Westerners hear of few beyond occasional earthquakes and the typhoons that roar in from the Bay of Bengal. An American metropolitan newspaper editor once told me that in placement of stories, one local death equals one thousand in India.

I had not planned on traveling to Andhra Pradesh, but this calamity troubled me. After I heard the story, in a hushed voice, from a professor in New Delhi, I could not stop thinking about farmers dying by their own hands on their own lands. In the days that followed, as I tried to learn more, I became convinced that in the roots of these farmer-suicides, in the fatal connection between farmers and their land, I could find lessons about the proposed marriage of biotechnology and the developing world. I began making my way to Warangal, where the trouble is the worst.

Accidents along the road remind me that life in India is a risky proposition. In a crash just in front of my hired car, a tour bus swerving to pass struck one of the ubiquitous, three-wheeler rickshaws' and sent it hurtling end over end, terrifyingly fast, like a punted football bouncing on turf. The bus didn't stop, and neither did my driver, Vascudine, who shook his head and mouthed quietly: "It is very dangerous. Everybody speeds up, and there is no control." Later, where traffic is slowed and diverted, I see a robin's-egg blue scooter stuffed under a bus. And blood. On India's roads, about seventy thousand people die every year and one million are seriously injured. I see why.

Preparing for this journey, I had talked about risk with Per Pinstrup-Andersen, who is executive director of the International Food Policy Research Institute. His organization, which is comfortably housed in downtown Washington, is one of sixteen in a global network ponderously named the Consultative Group on International Agriculture Research. It is funded by the Rockefeller Foundation among others, and it considers itself the preeminent alliance looking toward the future food needs of the world's poor.

Pinstrup-Andersen admonishes the life-science behemoths to share their genetic technologies. He is an influential supporter of biotechnology at a time when the industry needs friends. When I met him at the World Bank, a shimmering metal-and-glass edifice on Pennsylvania Avenue two blocks from the White House, Pinstrup-Anderson, a tall, bearded Dane, asserted that food security cannot be achieved without genetically engineering seeds. He said something else that I recalled on the road to Warangal: "The acceptable risk level for a poor person is very different from an acceptable risk level for a rich person."

Calculating risk and benefit is what the global debate over genetically modified food is about. For India, the potential risk of GMOs is especially real to those who worry about globalization and the threat to India's self-reliance. But when it comes to human health and the environment, the risk of genetically modified food seems, to an outsider, dwarfed by the risks in the water, on land, and in the air.

In New Delhi, two billion liters of sewage and waste flow into the Yamuna River every day. Most species of fish have long since disappeared, and migratory birds no longer stop here. In my days as an investigative reporter, I hunted dumps in developing countries to track shipments of toxic wastes from the North. In Mexico, in Central America, and as far away as South Africa, I would ask people to guide me to secret landfills and the repositories of midnight dumpers in my search for clues to the identities of predatory companies from industrialized countries and the waste brokers who rid them of their poisons. In India, little searching is required; open dumps and fetid piles of waste assault visitors. In the coming years, New Delhi is threatened with becoming an "overcrowded, powerless slum buried deep under garbage," the government's National Capital Regional Planning Board had warned.

Already, 225 million acres of land in India are degraded and barren from salt built up by farm irrigation. Much of the lost land is in the prime agriculture regions of the Indus and Ganges. That is ominous for a country of hungry people because neither rice nor any of India's other prime foods tolerate salty soil. Biotechnology's defenders score when they promise that one day, genetic engineers might succeed in designing plants and trees that thrive in salty and acidic soil.

In New Delhi, a city of thirteen million, roughly eight thousand five hundred people die every year from pollution, most of it in the air, the World Health Organization estimates. Countrywide, two million people develop tuberculosis annually and a half million die from it. In New Delhi and six other cities, over 50 percent of children under the age of twelve suffer from serious lead poisoning, a foundation study concluded.

In the storied Ganges River float decomposing human corpses cast from the banks in Hindu last rites. One remedy proposed in the 1980s was to release thirty thousand flesh-eating turtles. But the turtles couldn't keep up with the bodies—or avoid the poachers. Turtles are a level of technology that many in India want to get beyond.

THE FATAL CONNECTION

At dusk in Pegadapally near the city of Warangal, men and women arrive home from the fields leading oxen bearing faggots for their evening fires. Ileakamru Shankar already has a pot of water steaming on an open fire in front of his home. He motions for me to sit on a wooden bench outside the entrance. His wife, Masuri, returns with a framed, color photograph of their son, Damera.

At noon on the last Monday of the preceding February, Damera Shankar walked the red dirt road from the village to check his two-acre plot. For over an hour, he watched voracious pests, especially a caterpillar known hereabouts as *ladde purugu,* sucking and chewing his cotton plants. Around one-thirty that afternoon he returned to his home and, unbeknownst to his family, he mixed a cocktail of two insecticides. He tilted back his head and poured it down. Then he made his final trip along the red dirt road and lay down in the field that had betrayed him. At five-thirty, they found his body.

At age twenty-three, Damera Shankar already was thousands of dollars in debt for chemicals that no longer killed the pests. Seeing the ruin of his crops that day, Shankar believed his situation was hopeless and that his family would go hungry. He was shamed, his father surmises. "Until then, he had kept confidence. But when he saw his field"—Ileakamru's voice breaks—"he knew he had lost control."

I ask the father in a different way why his son had done this.

"He could not bear the burden of debt; he was feeling anxiety. The insecticide was not working and, also, I think, there was a problem with the seeds. But I could not understand this. I was not expecting it," the father replies, shaking his head slowly. Masuri Shankar, still holding the photograph, weeps.

In Central America a few years earlier, I'd stumbled on this phenomenon of desperate people succumbing to the technology meant to help them. In Central America, the poison of choice is paraquat, a potent herbicide that *campesinos* spray from tanks strapped to their backs to kill weeds. Investigating misuse of pesticides on Guatemala's southern coastal plain, I had arrived at a hospital in Mazatenango just four hours after Oscar Lopez Calderon, seventeen, swallowed paraquat. On pink-striped sheets, while pressing his stomach with both hands, Lamaza responded faintly to me. "I don't know why I did it. I felt stress. The paraquat, it was there, and I . . . I just took it," he murmured, before looking away. When I phoned two months later, a nurse told me that Oscar had gone home from the hospital and was hanging on "in a lamentable condition."

The pesticide-suicides in Central America often were sparked by liquor, family disputes, or both. A farmer got roll-in-the-dust drunk on Saturday night and, while stumbling home, forced himself on the neighbor's widowed sister. Sober at daybreak, he realized the shame of it all and reached for the paraquat. In another Guatemalan hospital where I visited, an eighteen-year-old farmer told me that he drank paraquat after a fight with his wife. I talked to friends of a thirty-year-old man who suspected his wife of having an affair. He downed a liter of insecticide before he found out for sure.

In Andhra Pradesh, the farmers are dying for altogether different reasons. Almost always, the trigger is a cycle of pests, chemicals, crop failure, and debt. Outside Warangal, another victim, Gandreathi Bixapethi, who was twenty-eight, had accumulated a fortune in debt during crop failures unabated by chemicals. He brooded, growing estranged from his family. In a field on the first day of June 1998, they found his body alongside an empty liter bottle of endosulfan.

"He was a good son and a hard worker. He even would do the work of women," his mother tells me, speaking of the task of weeding.

Later, as we drink strong black tea on stools outside his home, Sykender Reddy, an unelected leader of his village, overlays the despair with the daily fight of the residents of Pegadapally to survive. "There are people around who can't get two meals a day," he says. "It is very difficult to make our dreams."

Reddy knows that genetically engineered crops are being tested in his country, and he has heard of their capacity to ward off insects and even to help grow more food. He said farmers must consider GMOs. But as he weighs their promise, an instinctual distrust of outsiders pokes through.

"It might be welcome," he says. "But if it is in the hands of the

In a village in rural India, Masuri Shankar holds a photograph of her son, Damera, among the hundreds of Indian farmers who committed suicide in desperation over farm pests.

multinationals, they could make the price high and rule over us. It will be a difficult decision to make."

REVOLUTIONARY BACKLASH

India has been an agriculture testing ground before: in the Green Revolution, which enabled India to increase wheat production dramatically through hybrid seeds and fertilizers. In 1964, ships docked in Calcutta delivering high-yielding Mexican wheat and fertilizers. The wheat and the rice that followed increased the grain yield in India, saving the lives of tens of millions of people.

Among the dozens of scientists, farmers, and government officials I spoke to in India, only a handful dismissed the importance of the Green

Revolution in the country's history. But some of the farmers and a segment of the country's radical thinkers told me they are feeling betrayed by the Green Revolution's promise. Accompanying those shipments of grain were the seeds of a new system of intensive farming, reliant on fertilizers and pesticides, that spread across the country. The new ways of farming have enabled India to grow more food. But they have left many farmers on a treadmill of chemicals and debt. Now, with suicides as a backdrop, many farmers are wondering what next the foreign companies have in store for them.

A. Sudershan Reddy, an economics professor at the local college in Warangal, has assembled the dates and the details of many of the suicides in a 135-page, footnoted report that he calls "Gathering Agrarian Crisis." By April in Andhra Pradesh, the tally of deaths since the previous autumn was 350, which he had broken down with numbers on a map. He pulls out a notebook and, by month, we write down the suicides reported to him in the Warangal district alone since: thirteen in June; eight in July; twelve in each of the months of July, August, and September; six in October, and twelve already in November. It comes to 425, not counting the rest of Andhra Pradesh.

Reddy and his colleagues painstakingly tracked the tragedies through records in villages, phone calls, and personal visits. He says that his tally is unofficial, but it is unlikely that an official death count would ever come from a government ashamed of what has transpired.

When he greets me at his home in Warangal, Reddy, a slight man just over five feet tall, is wearing a starched, white-linen suit and brown sandals. Twice during our conversation, he utters these words: "What we have here is a grave situation, a gathering agrarian crisis." He gives me his analysis of the suicides and asks that I make copies of it when I return to Washington and give them to agricultural economists who, in his view, devote too much academic energy to trumpeting the success of the Green Revolution.

Reddy disputes those who argue that the suicides in India have been triggered solely by back-to-back seasons of freakish conditions—untimely rains followed by drought—abetted by the risky decision to grow cotton rather than food crops. In his studies, he found that only one-third of the farmers had grown cotton exclusively; most who had died also grew rice, corn, and chilies.

In talking to the families of fifty of the suicide victims, Reddy and his students found similarities. The dead men were young: 75 percent under the age of forty; 80 percent held under five acres of land. Their average debt was 52,000 rupees, or just under $1,500. One farmer had a debt of $7,140; another owed just $142. Often, the ill-fated farmers had grown estranged from their families, too ashamed to talk about their cascading problems. They needed counsel but, Reddy found, they had no one to guide them. Nearly all of the farmers on Reddy's list had fallen into debt buying pesticides, which were easily obtainable on credit from hundreds of pesticide shops in Andhra Pradesh.

In India, often the pesticide sellers and agricultural extension advisors are one and the same. To illiterate farmers, these salesmen-advisors peddle chemicals that have fallen behind the insects' evolutionary cycle. That happens quickly with cotton, a fickle crop to begin with, and one especially exploited by the pestiferous American bollworm. The ubiquitous sellers of chemicals, viewed as a scourge from the Green Revolution, feed the cycle of chemicals and debt.

The use of pesticides has become mindless, Reddy says; farmers will apply them as many as twenty-five times during a growing season whether or not they have seen bollworms, white flies, or other pests. "They have invested so much in a crop, they feel they must take every precaution," he said.

As chemicals are deployed with ever-increasing frequency, they saturate the land, killing beneficial soil organisms. A separate study I saw tells another legacy of abuse: 72 percent of rural women in India have DDT in their breast milk, even though DDT has been banned in the country.

After what he has seen, Reddy is not generous toward the companies that produce the chemicals. Nor is he hopeful that genetically modified plants will ease the plight of farmers in the future. "Food security is dangerous if you depend on multinational companies," he tells me. "Their goal is to make profit."

As I leave Reddy's home, he sums up the suicides: "The pesticides are supposed to kill the pests, but they're killing the farmers instead," he says.

A MONSANTO SOLUTION

Several days later, in the city of Bangalore, I watch the squirming American bollworm which, for American interests, is unfortunately named; it did not originate in the United States, nor does it afflict American farms. Here in Bangalore, the bollworms and other pests are confined to the laboratory of Monsanto, which is working toward genetically engineered remedies. Recently, Monsanto has taken its technology into forty one-acre plots in India to test varieties of its Bollgard cotton, modified with the Bt gene—which results in the plants producing a protein that, when ingested by pests, acts like an insecticide.

Over a lunch of rice, curry, and yogurt, T.M. Manjunath, a Monsanto entomologist who later would become director of the lab, tells me: "People say 'Americans gave us this pest and now they're hoping to sell us another means to control it.' If using transgenic seeds would mean spraying two or three times rather than twelve or fifteen, it would be a great contribution."

In India, where two-thirds of the population lives on 106 million farms, Monsanto sees a vast market for its chemicals and genetic technologies. At the time of my arrival, Monsanto is moving swiftly to beat its life-science competitors into that market. Monsanto has increased its India staff tenfold and developed the means to distribute its technology by acquiring the overseas seed operations of Cargill and joining forces with Maharashtra Hybrid Seed Company, better known as Mahyco, an Indian seed company.

Monsanto planned applications to begin work on genetically modified rice, corn, and sugarcane, I am told. If Monsanto wins commercial approval for these crops, the company will establish control of modified seed markets for one-quarter of the world's farmers. In anticipation of such rewards, Monsanto hired part-time farm specialists to drive the countryside with the message that the company's new products can help.

"The key is getting in touch with the millions of farmers that have less than two hectares of land in this country. The potential ability to help in this country is just astronomical," Judith Chambers, a Monsanto international specialist from Washington, tells me when we meet in New Delhi.

To win over Indian government officials, Monsanto has flown them to the United States. When I asked P.K. Ghosh, from India's Department of Biotechnology, about his trip to St. Louis, he responded, "My God, what they have there is nice."

In the spring of 1998, Monsanto opened its research laboratory and forged a partnership with the respected Indian Institute of Science. Still, resistance to the company's research has been as determined as India's pests. "Ever since the lab opened, it has been in the news as if Monsanto had committed some sort of crime," Monsanto's Manjunath says.

Monsanto has heard the same threats I have heard from Nanjundaswamy's organization in the state of Karnataka. Manjunath is trying to head off trouble. His daughter and the daughter of Nanjundaswamy, the organization's leader, attend the same school. Nanjundaswamy's daughter is studying science, and Manjunath stopped by the Nanjundaswamy home to invite father and daughter to Monsanto's new laboratory. They hadn't yet paid a visit.

"They don't understand science. That is a tragedy," Manjunath says. "I firmly believe that it is only the transgenic crops that can contribute to productivity and food security in India."

Meanwhile, Monsanto was taking precautions. In its new, eight-thousand-square-foot greenhouse built in Bangalore, the genetically engineered crops that the company says will feed India's hungry people and relieve the misery of pests sprout behind half-inch-thick, bulletproof plastic.

THE OPPOSITION

The Violence of the Green Revolution. Biopiracy. The Future of Our Seeds. MONSANTO: Peddling 'Life Sciences' or 'Death Sciences.'

The books and pamphlets so titled are authored by Vandana Shiva and on display at her Research Foundation for Science, Technology and Ecology, situated in a beige-brick building in the clean and comfortable Hauz Khas section of New Delhi. Besides books, the shelves are lined with jars of traditional Indian rice varieties and seeds, among them mustard.

The nondescript center is the base of operations for Vandana Shiva, intellectual, feminist, and a Bengal tiger on the attack against biotechnology in her native land and around the world. She has been called the first international rock star of the environmental movement, and after watching her in action, I have come to understand her allure. She is exotic to Westerners in her flowing orange saris, and her magnetic smile belies the ferocity of her beliefs. She relishes matching wits with whomever she's pitted against. She's skilled at revving crowds with persuasive, if simplistic views. When critics of genetic engineering gather, especially in Europe and more recently, in the United States, Shiva is sought out as the prime-time talent, the motivator. She told me that she spends one-third of her time traveling around the world, one-third overseeing her New Delhi foundation, and the remainder in what she termed "restorative pursuits" at her family farm in the Himalayan foothills.

"Living resources should not be bought and sold. They are not like machines," I have heard her say. "Genetic engineering is a technology born in the commercial sector rather than in universities and in public debates."

Shiva, born in 1952, has a scientific background and a Western education: In Canada, she earned a master's degree in physics and a doctorate in the philosophy of science. She is a prolific writer and organizer who lays the foundations of arguments echoed by anti-GMO forces around the world. She nurtures her alliances through the Internet.

Shiva's thinking reflects the potent nationalist sentiments in her country even though she has been aligned with the more moderate Congress Party. Her influence extends beyond India's borders. It was on display in 1998, when she helped torpedo an agreement between Monsanto and the Grameen Bank. The bank and its founder, Muhammad Yunus, win praise in lending circles for risky and trail-blazing microcredit, noncollateralized loans that enable small farmers to obtain credit. Monsanto agreed to provide $150,000 to the bank in a partnership that would establish the Grameen Monsanto Center for Environment-Friendly Technologies in Dhaka, Bangladesh. When the agreement was announced, Monsanto observed that the "thousands of poor Bangladeshi weavers working with handlooms" stood to benefit.

To Shiva, the agreement was part of Monsanto's drive to expand its empire, a marketing ploy to open markets for transgenic seeds. The campaign forced Yunus to back out of the deal and infuriated officials at Monsanto. It's one more reason Shiva is reviled by some in her country's Western-leaning agricultural establishment.

In her view of the world, genetic engineering is a means of control bordering on enslavement. "I see the pushing of genetic engineering and the collusion between Monsanto and the U.S. administration as the new imperialism," she told me when we met in Ireland, in 1998.

In St. Louis, where she helped to lead a protest at Monsanto's world headquarters, she asserted that genetic engineering will enable "a new level of corporate control over agriculture around the world."

Shiva regards the Green Revolution as a "myth." India and Pakistan would have survived without the Western technologies, she has said. She put that view in a historical context in *Bija,* a quarterly magazine she edits, in 1998: "For ten thousand years, farmers and peasants had produced their own seeds, on their own land, selecting the best seeds, storing them, replanting them, and letting nature take its course in the renewal and enrichment of life. With the Green Revolution, peasants were no longer to be custodians of the common genetic heritage through the storage and preservation of grain. The 'miracle seeds' of the Green Revolution transformed this common genetic heritage into private property, protected by patents and intellectual property rights."

The suicide epidemic is, to her, further evidence that India's farmers have lost control over their seeds, their agriculture, and their knowledge, falling prey to the allure of growing "white gold"—cotton—rather than food. They "have switched to the mentality of technological fixes to ecological problems," she says.

Shiva placed the number of farmer suicides in her country at twenty-five thousand over three years, when she appeared in the summer of 2000 before a congressional panel studying hunger.

DELICATE BALANCE

Shiva is not universally read in her own land, for half its population is illiterate. Many of the farmers I spoke with in villages and fields had not heard of Shiva or genetically modified crops. Several I met in the south India village of Hoskote were of the same mind as Krishna Murthy, whose sixty acres make him a wealthy Indian farmer. "If anything like that is out there, we'll put it to the test," Murthy said of the pending arrival of genetically modified seeds.

Unfortunately for Monsanto, the Terminator—the technology that renders seeds sterile in order to protect intellectual property—is the first introduction of many Indian farmers to the new technology. In a land where three-fourths or more farmers save their seeds, it embodies their fear: outside control of what they grow. During my visit, the *Indian Express* newspaper bannered a story about a farm exposition with the headline: "Ryots [farmers] will meet terminator at Kisan fair." Alongside, a photo showed farmers suspiciously eyeing a contraption with arms and hoses, which may or may not be a tractor. Feeling mischievous, I faxed it to a Monsanto official in Washington with the note, "I had to travel nine thousand miles to see what the Terminator looks like."

Around the world, opposition to GMOs divides into two general though related categories—grass-roots and political. In Europe, each form of opposition has fed the other. In India, despite the pockets of resistance and nationalist sentiments, I got the sense that neither opposition will be insurmountable. Already, three hundred million people in India don't get enough to eat. The future will compound that crisis. Every year, India adds 1.8 percent to its population; since the 1970s, India has swelled by as many people as live in the whole United States. By 2025, with twenty million new mouths to feed each year, India will need to produce sixty million more metric tons of grain yearly than it can produce in the year 2000. That's 30 percent more—to be wrung out of agricultural lands that continue to shrink as global warming swells the oceans, which in turn inundate once-arable acres with salt.

The promises of genetic engineering closely match the needs of India's farming majority: growing more food, stemming erosion, and lowering farmers' costs. In theory, genetic engineering is infinitely adaptable: In India, for example, genes for salt tolerance might be transferred from mangrove trees to rice and other food crops. Compared to traditional methods of breeding plants for selected traits, genetic engineering is fast as lightning.

But such feats are still, by and large, promises. I do not know when, or if, genetic engineering will increase the yield of food, and neither does the biotechnology industry. I do know that the Bt technology works to control pests, which means that, by reducing the need for chemicals, it could begin to remedy the chemical abuse that plagues India much like its pests. But there are risks: If pests develop resistance to Bt, then where will India's farmers be?

I am most hopeful for the technology in the hands of Indian researchers, especially those at the International Crops Research Institute for Semi-Arid Tropics. They are working with crops that receive little attention—sorghum, millet, groundnut, chickpea, cowpea, and pigeonpea—but feed tens of millions of people daily. For instance, scientists are developing sorghum plants that resist the striga parasite. They're putting Bt in pigeonpea, which gets chewed up by bollworms, and they are intent on incorporating foreign genes to overcome viruses and fungal pathogens.

On the one side, skeptics worry that the life-science multinationals will trample on their culture. On the other side of the balance, people poisoned by chemicals still go hungry. Weighing both sides, I think India would make a mistake to let these new seeds of hope wither on the vine. India has a reputation for its fine scientists and technicians; in a decade, its computer software industry has grown from $150 million to a $4-billion-industry exporting to the United States and around the world. Even with the uncertainties of biotechnology, India seems capable of controlling its technological future.

Considering India's future, I reflect on the ten incarnations of Vishnu, the Hindu god with the human face. In the fifth incarnation, Vishnu is a dwarf named Vamana. Bali, the tempter, tells Vishnu that he can have all the land he can cover in three strides of his pathetically short legs. Vishnu outsmarts Bali by changing himself into a giant who covers the countryside in three huge strides.

In modern India, pesticides are the demon Bali, and Vishnu awaits transformation.

"KNIFE UNSHEATHED"

After traveling India north to south and back again, I am writing my thoughts in my New Delhi hotel—until I accidentally kick the blue metal box I have borrowed to convert the electric current. It sizzles and pops, sending surges of current that explode two lightbulbs. One of the bulbs shoots upward like a missile and bursts against the ceiling. There is an acrid smell, and the room goes black—except for my laptop computer. It, too, is beeping an injury message. I have seconds, I know from experience, to sum up what I have seen in India.

As my mind grasps for a straw of wisdom, my fingers type the words of M.S. Swaminathan, the famous Indian geneticist from Madras, a winner of the coveted World Food Prize: "Any technology: You can use it for good or bad. If I have a knife, I can cut you—or I can cut your food."

Before all is black, I write my conclusion: "To India, genetic engineering is a knife unsheathed, able to cut either way."

EPILOGUE

Three weeks later, I am back in Washington, when among my e-mails pop up photos sent by M.D. Nanjundaswamy, the militant farm leader. When I open them, they show farmers crowded around a bright orange glow. It is not, as I used to see in the Midwest, a wiener roast.

The photos show dozens of farmers on what I'm told is a plot of genetically modified cotton in the Bellary district in the state of Karnataka. They had ripped the cotton plants from the soil and made a pile of what they had unearthed.

In Indian lore, Agni, a warrior riding a ram, blew flames from his mouth in battle. The farmers of Karnataka Rajya Raitha Sangh used gasoline and a match, as the flames rising from the genetically engineered cotton in the photo showed. Their attack, they said, was part of "Operation Cremation Monsanto."

They were not finished.

Six days later, in the Raichur district of Andhra Pradesh—where pesticides have forged the fatal connection between farmers and their land—agitated farmers assembled at another plot of cotton planted in seeds girded to resist pests. At first, the landowner who had struck the bargain with Monsanto resisted. But Nanjundaswamy talked with him, and the man retired silently as the mob lit their fires again.

I am less certain after seeing the pictures that the magic seeds will take root across India anytime soon.

PLANTINGS NINE

I've become fond of watching my genetically engineered soybeans grow. I see why the Chinese called them "Yellow Jewel," "Great Treasure" or, my favorite, "Brings Happiness."

In China, cultivated soybeans were sprouting more than three thousand years ago. Long before genetic engineering, they were bred to stand about three feet tall and just as wide. In the United States, we didn't have the sense to plant soybeans until the eighteenth century. They grew in obscurity for more than a century, even though Ben Franklin had trumpeted them as marvelous for vegetarians. In the early 1900s, Americans figured out how to crush soybeans for oil. Almost overnight, the world became dependent on "Heaven's Bird," another name for the soybean, and it's easy to see why.

With a protein content of 40 percent, along with vitamin E, calcium, potassium, and a half dozen anticarcinogen agents, surely they are as versatile a plant as ever popped from the soil. These days, soybeans are planted on seventy million acres in the United States; no country produces them as bountifully.

I could eat my beans after boiling or roasting them (they taste like limas). I could go the route of the ancient harvesters and make miso, soy sauce, or tofu. Or I could get ultramodern and try to convert them into tofutti ice cream. I'm not sure what I'll do when my experiment concludes. I asked the American Soybean Association for advice (I didn't tell them I was growing modified beans illegally), and they gave me a list that, single-spaced, would be taller than my knee-high plants.

I could turn them into animal feed; baby formula; candy; cereal; doughnut mix; margarine; mayonnaise; noodles; pancake flour; sausage casings; soynut butter; soy coffee; soy milk; or shortening. The list gets stranger: adhesives; asphalt; cosmetics; fire-extinguisher foam; ink; insecticides; linoleum; magnetic tape, paint; paper; particle board; plastics; putty;

rubber; shampoo; soap; soy diesel; varnishes; waterproof cement; and, for subway use, graffiti remover.

I'll have no problem finding a use for my bucket or two of beans, and I won't need a combine to harvest them. Shortly, I'll just crawl in the dirt as I did when I planted them.

18

BIOTECH AND THE PARADOX OF PLENTY

Whoever could make two ears of corn, or two blades of grass grow upon
a spot of ground where only one grew before, would deserve better of
mankind, and do more essential service for his country than the whole
race of politicians put together.

—The King of Brobdingnag; *Gulliver's Travels*, 1727.

FLORENCE WAMBUGU IS a robust African who, when I met her, presented a queenly figure in a shimmery black dress and gold jewelry reflecting the Caribbean sun. Even far from her home, in Kenya, she exudes self-confidence, and she speaks with confidence of the bounty genetic engineering holds for her continent.

"I think that there is a great potential to increase food production in Africa with these technologies," she told me when we talked in Colombia.

It was early in 1998, and her organization, the International Service for the Acquisition of Agri-Biotech Applications, an industry-funded collaborative, had been sprouting genetically engineered sweet potatoes in a greenhouse in Kenya. Soon, she planned to supervise field trials with the bioengineered plants, sowing them outdoors in fifty-by-one-hundred-foot plots.

The sweet potato holds a prominence among foods I had not expected: It is the world's sixth-largest subsistence crop. Grown in South America in pre-Inca times, it made its way to Spain, and thence to Europe. European writers who referred to potatoes, Shakespeare among them, probably had the sweet potato in mind. Because it will withstand drought, heat, and thin soils, it is favored by the poorest of the world's farmers and those tilling unforgiving patches of earth. But, like many who grow them, these tropical tubers suffer disease and predation. A degenerative disease called the feathery mottle virus is a chief affliction.

Wambugu, who has a doctorate in biology, explained to me how the virus withers the leaves and stems of the sweet potato plant, not always killing it but destroying its capacity to generate underground vegetables. On the African continent, the yield of sweet potatoes is less than half the world's average. For many Africans, a poor potato crop is a life-threatening matter.

In 1992, Wambugu, then a scientist with the Kenyan Agricultural Research Institute, traveled to Monsanto for a two-year stay. In St. Louis, she worked with company scientists who were well on their way to engineering potatoes to resist other pests, especially the voracious purple-and-cream-striped Colorado potato beetle. With help from Monsanto and the U.S. Agency for International Development, Wambugu and another Kenyan, Daniel Maingi, worked on methods of introducing virus-resistance genes into sweet potatoes and regenerating plants from African varieties.

These are tools, she insisted, that one day would help the people in her country who need help the most. "It's in the seed; you don't have to have other inputs to get the benefits. Even a grandmother who can't read or write can still plant the seed. To many people, when they see transgenic, it looks like a monster. To us, it is hope. Because we are prepared."

Near the turn of the millennium, the Kenya Biosafety Council bestowed its blessing on Wambugu's project. In mock-trials, the Kenya Agricultural Research Institute planted the potatoes in Kakamega, Kisii, Muguga, Mtwapa, and Embu. Finally, in August 2000, the Kenyan government formally declared that it was commercially launching its first genetically engineered crop, the disease-resistant potato. Among those on hand in Nairobi were Monsanto officials and erstwhile United Nations Ambassador Andrew Young, who, along with his Atlanta-based Goodworks International, was hired by Monsanto as an ambassador for biotechnology.

"Africa is on the verge of a tremendous revolution," Young proclaimed.

ROOTS OF HUNGER

In the debate over biotechnology, no issue is more provocative than the potential of genetic engineering to feed the world. It is the trump card slapped on the table by creators of the technology, played to shame any who would suggest that the risks of modified food remain unknown. *Who are YOU to stand in the path of feeding the hungry one day?* they ask the environmentalists and the consumer advocates who hold up flags of caution.

To the skeptics, the shame lies in invoking the needs of the hungry who—save for a handful of charitable efforts like the Monsanto alliance with Kenya—have gained little from a technology tailored to the commodity crops of the vast fields of the North. Never has it been the mission of multinationals to feed the world, biotechnology's critics argue, and there is little scientific data today showing yield gains from newly spliced seeds. *So how can YOU be so cynical as to offer unfounded hope in order to sell your products?* the critics ask the creators.

Feeding the hungry is the Final Argument, the clash that trips the emotions on both sides in the biotech debate and keeps them a chasm apart. It's a debate that branches out like the roots of tropical tubers into other systems of farming: polyculture versus monoculture, intensive versus traditional or organic. In the middle are the relief agencies, foundations, and earnest folks who seek solutions and lament the divisiveness of biotech squabbles and the energy squandered. It's a debate that plays out differently around the world: Europeans don't readily accept the assertion that hunger can be vanquished by increasing production. The average North American, on the other hand, is swayed by the notion that abundance is the ticket to relieving hunger's misery.

When either side talks of feeding the world, they're referring primarily to sub-Saharan Africa, India, Pakistan, and Bangladesh. The United Nations estimates that early in this new century, the world has 840 million chronically hungry people, primarily in these lands. That statistic will rise as the world's population increases from six to eight billion in 2025 and to nearly nine and a half billion by 2050, the United Nations predicts.

Hunger has its roots in poverty and landlessness, traveling mates in developing lands. Here, agreement can be reached. The conflict begins when the subject turns to food production. The biotechnology companies and their supporters argue that the equation for the future must include growing more food, much more, from roughly the same 1.7 billion acres of arable lands, where productivity has plateaued. The way, life-science companies and their allies say, is helping to increase production by controlling pests, stemming erosion, and lowering farmers' costs—and later engineering nutrition into plants.

"Worrying about starving future generations won't feed them. Food biotechnology will," asserted a now-famous Monsanto ad that first ran in 1998.

If Malthus and Mendel could meet, now would be the time. Thomas Malthus, unless he has changed his tune, would argue that, finally, the population is growing at such a rate that the denizens of this earth will no longer have the means of subsistence to keep pace. Don't worry, Gregor Mendel might say; my successors have figured out solutions. Lacking that historic meeting, the most enlightening debate I've heard on the subject occurred on Capitol Hill in the summer of 2000. Until then, members of Congress had not taken time to overlay biotechnology on the issue of hunger.

Starring in the discussion was "golden rice"—the beta-carotene-enriched variety that received global attention in the preceding months—so it was fitting that the Congressional Hunger Center's briefing took place with standing-room only in the Gold Room of the Rayburn House Office Building. Participants (in order of appearance) were: the Reverend David Beckmann, president, Bread for the World; U.S. Congressman Tony Hall, a Democrat from Ohio; Senator Richard Lugar, a Republican from Indiana; U.S. Congressman Dennis Kucinich, a Democrat from Ohio; Martina McGloughlin, professor, University of California at Davis; Vandana Shiva, of India, author and director of the Foundation for Science, Technology, and Natural Resources; C.S. Prakash, professor, Tuskegee University; Michael Hansen, research associate, Consumers Union Policy Institute; Per Pinstrup-Andersen, director,

International Food Policy Research Institute; Arthur Getz, World Resources Institute; Peggy Lemaux, University of California at Berkeley; Michael Pollan, author.

Here is some of what was said:

David Beckmann: Can biotech help?

"There are something like eight hundred million people in the world who are so poor that they don't get enough to eat. They lack energy. They are often sick. Their kids die in large numbers. They are tremendously vulnerable. They suffer abuse from virtually everybody above them in society. I am convinced that it is quite feasible to reduce the number of hungry people in our world to about four hundred million by, say, 2015. Despite the population explosion, there are fewer hungry people in the world today than there were twenty-five years ago. . . . Can biotechnology help fight world hunger?"

On Capitol Hill, Congress typically gets the first word, if not the last. To begin this gathering, two House members and a senator from the American heartland brought answers to Beckmann's question. As chairman of the Select Committee on Hunger, Tony Hall, a born-again Christian from Dayton, Ohio, has traveled to the world's hungriest lands: Haiti, Somalia, Uganda, Rwanda, and North Korea. Senator Richard Lugar of Indiana, Eagle Scout, Rhodes Scholar, and erstwhile presidential aspirant, has been one of Capitol Hill's most influential voices on farm issues. He would be followed by Representative Dennis Kucinich, a fearless, left-leaning Democrat from Cleveland who was making a mark as biotech's foremost critic in Congress.

Congressman Tony Hall: In need of political, spiritual will

"Biotechnology might be a big 'if.' Some people say there is very little in it, that it's irrelevant to the poor and hungry people, and its potential is uncertain. [They say] if this is about making money, I am not interested. So many things on Capitol Hill are about making money.

"If this is about hungry kids, I am interested. We would be foolish not to see a future role for biotechnology in helping them. I don't want to be remembered as the guy who predicted a world market for maybe five users, as an IBM chairman did in 1943. I am interested in what scientists are learning about biotechnology, and I would be wary of efforts to stifle their work. . . . We need more political will. We need more education of people in our own country and overseas. We need more spiritual will."

Senator Richard Lugar: As when the iceman cometh

"I begin with a tale of technology, theology, and ice. The revolutionary idea that ice from colder climates could be collected, transported to lower latitudes, stored, and eventually used as refrigeration was the brainchild of Frederick Tudor. In 1833, Boston's ice king demonstrated the feasibility of ice for refrigeration in dramatic fashion, moving two hundred tons packed with sawdust aboard a ship bound for India. For 180 days at sea, through mostly warm waters, the ship arrived and successfully delivered half of its original cargo. Tudor drew considerable criticism from prominent theologians, who argued that keeping ice underground in summer, similar to the practice of raising flowers under glass in winter, reversed the natural order of the universe and was, therefore, sinful. Nonetheless, the benefits of technology, allowing for extended storage of meat, fruits, and vegetables, were soon apparent.

"The rise and fall of ice delivery and the storage industry serves to illustrate three points that I believe are worth remembering: First, opposition frequently accompanies technological innovation. Opposition arises from the fact, myth, or cherished belief, and the obvious difficulty is to determine an elusive truth. Secondly, technologies that eventually win acceptance do so after demonstrating a clear benefit to society with few risks. A ready supply of ice fundamentally improved the safety and dynamics of food distribution, reducing disease and infection, especially for those living in America's expanding cities. . . . Innovation often provides fixes for earlier deficiencies but in the process may lead to a different set of concerns.

"Agricultural biotechnology is not unlike Frederick Tudor's ship leaving Boston Harbor with a cargo of ice. On the docks, a crowd quickly gathers, split between voices offering encouragement and voices offering disapproval. My fear is not that agricultural biotechnology has inspired controversy, but rather the debate has become polarized, sometimes reactionary, so as to preclude reasoned public debate over the merits of new technology versus possible risks. . . . If staple foods that these poorest of poor children eat each day could be fortified with vitamin A through the application of biotechnology, a worldwide scourge of blindness from dietary deficiencies could be alleviated. . . . Demographers predict that the population of the U.S. will double over the next hundred years and rural population will increase by fifty percent by the year 2050. Development and the need for housing will place an inexorable pressure on land that now constitutes a significant percentage of America's and the world's treasured open spaces. . . . If agricultural efficiency remains static, then more land will be needed to grow more food.

"Faced with the choice of starvation or cutting down the rain forest, mankind will have few options. But this is a fool's game. . . . If developed with intent to improve the lives of people everywhere, biotechnology can increase agricultural efficiency, reduce chemical pesticides, and improve food's nutritional value. It is a difficult public challenge."

Congressman Dennis Kucinich: A "powerful illusion"

"Is biotechnology and its derivative, genetically engineered food, the solution to solving world hunger? The answer is no. For the world is a cornucopia of food, yet people are still hungry in all nations, including this one. This paradox needs to be examined. Is it possible that people go hungry because of political obstacles and severe economic hardship? We all know that even here in America, in this country of plenty, many American families go to bed each night hungry, some because they can't pay for dinner. Perhaps a living wage would help them. In many less developed nations, financial hardship, poor distribution of food, and political causes are the most troubling issues to face.

"Will biotechnology help alleviate these causes? Perhaps sustainable agriculture and sustainable economic development will help feed these people. The so-called Green Revolution was supposed to solve world hunger. It didn't. Biotechnology offers a similarly powerful illusion. If the biotechnology industry wants to reduce world hunger, they can certainly use their great resources to protect the indigenous natural biodiversity of developing nations and to encourage sustainable agriculture that does not require expensive agricultural inputs. . . . Even if genetically engineered food has some yet-to-be-discovered intrinsic benefit, this benefit certainly does not override the people's right to know and the necessary assurance that the food is safe.

"I believe that people in the less-developed nations also have the right to know what they eat. . . . We are our brothers' and sisters' keepers. The New Testament asks 'Who among us, when our brother asks for a loaf of bread, would instead give him a stone?' We must answer many questions before we can safely assume that the wonderful instinct that we have to feed the hungry is a true fulfillment of a spiritual mission when we feed the hungry genetically engineered food."

Hunger is an affliction that few Americans have seen, let alone experienced firsthand. Likewise, the United States has only begun to witness the ferocious debate over biotech that has exploded around the world. In this gathering, organizers have called on the expertise of foreign-born citizens and experts from countries where the issue of genetic engineering sizzled on a front burner. Martina McGloughlin spices her defense of biotechnology with stories of growing up on a farm in Ireland. Vandana Shiva and C.S. Prakash, both India-born, always can be counted on to offer starkly opposing opinions about biotech's capacity to feed the world.

Martina McGloughlin: Starvation, not biotech, the enemy

"I spent my formative years on a farm, on my hands and knees weeding and, would you believe it, sowing and picking potatoes. I've really learned this process. My father always said if something doesn't kill you, it will make you stronger. I believe that there are better ways to build character than to have to scramble in the dirt.

"The human population continues to grow while arable land is a finite quantity. In fact, it's estimated that in fifty years, the amount of arable land will be reduced by half, so we must make optimum use of all tools to improve productivity and food production. Many scientists believe that biotechnology could raise overall crop productivity in developing countries as much as twenty-five percent and help prevent the loss of those crops after they are harvested. . . . We will be able to use biotechnology to enhance nutritional content of crops such as protein, vitamins, minerals, and antioxidants, remove antinutrients, remove allergens, and remove toxins. We will also be able to enhance other characteristics such as growing seasons, stress tolerance, yields, geographic distribution, disease resistance, shelf life, and other properties. The ability to manipulate plant nutritional content has the potential to directly benefit developing countries. Scientists can use similar plant delivery systems to provide not just intense nutrition but also vaccines and therapeutics that are especially important in developing countries.

"Stresses caused by pests, diseases, and harsh environments cause enormous amounts of crop losses in developing countries. . . . In Hawaii, the papaya industry was down to its last stand, destroyed by a tiny killer called papaya ring spot virus for which there is no natural resistance. A simple gene from the virus itself acted like a vaccine to completely protect the plant and restore the economy. . . . By reducing dependency on chemicals and tillage through the development of natural fertilizers and pest-protected plants, biotechnology has the potential to conserve natural resources, prevent soil erosion, and improve environment quality. . . . Environmental stresses such as drought, heat, cold, and nonoptimal conditions can also be addressed. . . . Yield is also an issue. By engineering metabolic pathways, we can greatly increase productivity by bypassing the physiological barriers that cannot be addressed using traditional crop breeding.

"The view that genetically modified organisms pose new or greater dangers to the environment for human health is not supported by the weight of scientific research. . . . Millions of people have already eaten the products of genetic engineering, and no ill effects have been demonstrated. If we abandon the scientific process in judging the safety of food, we will slow or destroy the advantages that will reduce the use of unsafe chemicals and agricultural practices, and we will limit the incredible potential for improved nutrition and quality that promises to strengthen the agricultural

New Delhi scientist and activist Vandana Shiva travels the globe with the message that genetically modified crops are dangerous to farmers in the developing world.

economies around the world. As President Jimmy Carter said, responsible biotechnology is not the enemy. Starvation is."

Vandana Shiva: Monoculture's threat

"My history has been the opposite of my predecessor; I have walked from academia to the farm. I decided to apply my scientific training as a physicist to agriculture because of how repeatedly I have seen facts distorted. . . . The environmental claim and the food-security claim are totally false. . . . All that is wonderful public relations, but there is neither the science nor technology to deliver those applications. They have been talked about for twenty-five years, and we don't yet have nitrogen-fixing genetically engineered plants because nitrogen fixing is not a single-gene trait.

"The only way we defend our soil fertility and protect our topsoil is by having cover crops, by having polycultures. Wiping out polycultures through breeding crops that are resistant to herbicides is not just a threat to the biodiversity that is being wiped out; it's a threat to the soil, and it's a threat to food security. . . . In my region in the hills, with very, very shallow soils and terrace farming, [we produce] six times more overall food yield than the intensive Green Revolution areas in the Punjab because Third World farmers don't grow monocultures, they grow polycultures. In Nigeria the home gardens cultivated by women on 2 percent of the land provide 50 percent of the nutrition.

"It is not true that without genetic engineering the world will starve. It is definitely true that in the trials assessed in this country, there is no yield gain; in fact there is a yield drag. . . . Hunger is not just about the quantity of food available in the world, of which there will not be more if we do genetic engineering in any case. Hunger is a destruction of entitlements, of people not having purchasing power, and purchasing power is collapsing around the world as agricultural systems push farmers to spend on inputs and get less for what they grow. . . .

"Farmers are getting nothing back. I would like to conclude with the Canadian Farmers Union's submission to their parliament. It says that while the farmers are growing cereal grains—wheat, oats, and corn—and earning negative returns and are pushed close to bankruptcy, the companies that make breakfast cereals reap huge profits. . . . Maybe farmers are making too little because others are making too much. That crisis is going to be aggravated with genetic engineering and biotechnology. . . . Genetic engineering is just too monocultural, too impoverished, too nonsustainable to be our bet for feeding the hungry."

C.S. Prakash: "Imperialism" toward poor

"Hunger is a disease, and there's only one medicine for that, which is food. We could either produce food or we could buy the food. Producing and buying food go hand in hand for 80 percent of the people in India and in most developing countries who are engaged in farming. As an agricultural scientist, my research has been in how we can produce more food and better food and how we can develop improved varieties of crop lines. My heroes when I started studying agriculture were the great environmentalists, Norman Borlaug and M.S. Swaminathan.

"Using genetics helped my country save so much valuable land from being under the plow and improved production tremendously when population was increasing by leaps and bounds. Our food production and the application of knowledge helped improve food production in a very significant manner. We were producing twelve million tons of wheat in 1960. Now we produce seventy-five million tons.

"Biotechnology is not going to solve all the world's problems, and it would be foolhardy to even talk about that. What biotechnology can do is to help develop better varieties of crops. Genetic modification is not new. We started genetically modifying plants when we walked out of the caves in the Stone Age. . . . In the past hundred years, using the process of hybridization, wide crosses, and even more brutal techniques such as irradiation, we have been developing newer varieties. Biotechnology is just one other tool that we have, one that is far better, more precise; it is a scalpel that we're using compared to the sledgehammer.

"So why is there opposition to biotechnology? We see a lot of rhetoric. One of the things I want you to understand is that the same people who are critical of biotechnology and ones who are critical of the Green Revolution are critical of a lot of things. . . . There's also this imperialist attitude that we somehow must keep Third World farmers from the clutches of this new knowledge, this Western knowledge, this imperialism and capitalism. I'm frankly sick and tired of hearing those kinds of arguments because I grew up seeing what local knowledge is. It's losing one-third of your children before they hit the age of three. Is that the local knowledge that you want to keep reinforcing and keep perpetuating?"

Apostles of biotechnology cheered the discovery of "golden rice," a variety enriched by genetic engineering with beta-carotene, which is converted by the body into vitamin A. It was years from the field. But finally, biotech had gone a step further than promising to feed the world's poor. There was something new to talk about in gatherings such as these, as we saw when Shiva and Prakash clashed over the true value of sprouting a better rice.

Shiva: "The language of giving away to the Third World hides a process
that takes place before that, which is the process of biopiracy, the process

of taking genetic resources, very often patenting them, and then talking of giving away for free a patented genome that is private property. Secondly, on the particular issue of vitamin-A rice, we have very simple alternatives to it. Just in the Indian state of Bengal, one-hundred-and-fifty greens which are rich in vitamin A are eaten and grown by the women. We don't have to wait."

Prakash: "I don't see any incompatibility between eating all those vegetables and vitamin-A rice. Having some weeds rich in vitamin A is not the reason to prevent this golden rice. We're not saying this will work; we are saying this is a technology that we think is appealing and let's give it a chance.

"Regarding biopiracy, I hear that rhetoric all the time. For my forefathers who stayed in India, I say thank God for biopirates. We are able to enjoy wheat, peanuts and apples, and everything else—the chilies that Indians love—that came from outside. Genetic materials traditionally have moved from their places of origin to other places. This xenophobic mentality that somehow everybody is stealing all of our genetic material flies against the development of science and technology that is very badly needed."

Next, the Consumer's Union Michael Hanson returned to the paradox of plenty.

Michael Hansen: Confronting "paradox of plenty"

"During the Green Revolution, what you saw displaced was rice-cropping systems where farmers would rotate rice with other vegetable crops. They also would have fish, frogs, and crabs in the rice fields, so they were harvesting not only the rice but green vegetables and proteins. With the Green Revolution, we saw an increase in monoculture of rice and you basically saw all these other things disappearing. Because the chemicals that were used in these rice systems killed fish, frogs, and other things.

"Partly this vitamin A deficiency, which is actually an indication of poverty, is coming because the diet has been simplified further and further. There is this paradox of plenty. We have eight hundred million hungry people, and yet we have more food per person in the world than at any time in our history. There is an average of 4.3 pounds per person per day. That is 2.5 pounds of grain, nuts, and beans, a pound of milk, meat, and eggs and about a pound of fruit and vegetables. So there is clearly enough to feed everybody. The problem, it seems to me, is distribution. How can you have increasing numbers of starving people when there is so much production? . . . If you survey these peasant and farm organizations in the South, maybe some would like forms of genetic engineering. But when I talk to them, they say one of the major problems is land reform."

As director of the International Food Policy Research Institute, Per Pinstrup-Andersen is not shy about lecturing both sides in this debate if they stray too deeply for his liking into the politics of biotechnology.

Per Pinstrup-Andersen: A question of morality

"It is not a matter of politics versus technology. . . . We need to talk about statistics, but we also need to talk about real people. We need to talk about the woman in West Africa living on one-half acre, maybe an acre of land, trying to feed five children in the face of recurrent insect attacks and recurrent plant diseases. For her, losing a crop may be losing a child.

"Now how can we sit here debating whether she should have access to a drought-tolerant crop variety? I don't understand that. She can reject it, but let's not block her access. It is her final choice if she has access to it.

"None of us at this table has the ethical right to force a particular technology upon anybody, but neither do we have the ethical right to block access to it. The poor farmer in West Africa doesn't have any time for philosophical arguments as to whether it should be organic farming or fertilizers or genetically modified food. She is trying to feed her children. Let's help her by giving her access to all the options and she can choose which parts of the solution she would like to have. . . . How can we sit here and say this part of the solution is unacceptable because we don't like genetic modification? We have no right to say that. Let's make the choices available to the people who have to take the consequences. Let them choose—and please don't stop the science that is needed to develop the cures for the diseases that I may someday suffer from."

The World Resource Institute's Arthur Getz reminded participants that they were on hand for a "remarkable moment": Never before had the Select Committee on Hunger publicly taken up the role of how biotech might or might not play in hungry lands. But Getz wanted to divert the discussion from science to the soil; from the conceptual realm to the hard lands where hunger persists, where soil fertility is marginal, and where people farm under stressed conditions.

"How we approach hunger is a production problem where the setting is really very risky; where the risks of crop failure are survival questions; where there is a real premium on a rapid return on investment; where there is a very strong aversion to purchased inputs," he said.

Getz then asked a question. "Is the technology one-size-fits-all or something that is locally adaptive?"

From Shiva and McGloughlin, he heard conflicting answers.

Shiva: Curse of the monoculture

"It is not true that industrial breeding, whether earlier in the Green Revolution or now with genetic engineering, increases land-use efficiency and helps conserve biodiversity better. Precisely by displacing diversity on the farm, it actually increases land pressure. . . . Basically, one hectare of a polyculture is producing what 1.62 hectares of a monoculture would produce. I think it is absolutely critical that each time we are told about yield increases removing pressure on land, we should ask the questions, what is that land producing now that will need to be grown somewhere else? Is it displacing wilderness, displacing biodiversity, or depriving people of nutrition, which is precisely what happened in India with the Green Revolution when protein crops like the parsleys and legumes disappeared to increase acreage on the rice and wheat. We have very severe protein malnutrition because the legumes that are a staple of everyone, including the poor, have in this period become a luxury of the rich.

"Finally, it is not at all the case that genetic engineering very quickly breeds anything. It doesn't do any breeding faster than conventional breeding. . . . The vitamin A rice everyone recognizes, the developmental work will still take the kind of time that it took for breeding of any rice variety."

McGloughlin: A precise, predictable science

"At this point in time, the main 'developing country' that is using genetically modified crops is China, and they have very rapidly adopted it. They are probably going to greatly help in increasing not only productivity, but in making China one of the dominant economic powerhouses of the world. . . . I can't agree with Dr. Shiva's notion that through genetic engineering you do not speed up breeding because when you insert these genes, using traditional breeding, you have to take years, up to fifteen years of crossbreeding and backcrossing, to get rid of all the traits you don't want. Using genetic engineering, it's very precise, very predictable, because you're taking single genes and introducing those very quickly into sustainable genomic background. The other thing is, using traditional agricultural slash-and-burn systems, you are definitely reducing biodiversity. You are increasing leaching of soil nutrients and definitely are increasing erosion."

Later, Peggy Lemaux, a plant biologist and cooperative extension specialist at the University of California, Berkeley, who was speaking for the first time, felt too much conflict in the air.

Lemaux: "Unsettling"

"This morning has been unsettling to me because it is a bit like watching a tennis match where you go, 'boing, boing,' 'yes, no,' 'right, wrong.' It is not a black-and-white issue. It is a gray issue. In my heart, I am a consensus builder. I don't believe there is going to be unbridled use of biotechnology across the world to address everything, nor do I think that is the right way to go. Nor do I think it's going to solve all the problems.

"There are places where it can be very effective. Do I think the public sector has a role in this? Very definitely. I think we have a responsibility as public-sector scientists to be involved in this technically and to be involved in the debate, and to ask questions as to how to move forward. . . . I think more than money, we need a structure that will provide us the opportunity, as scientists in a developed country, to find out what those issues are and where it makes sense to use these technologies or use organic approaches.

"I think sitting here today, probably none of us in this room, maybe there are a few exceptions, have really experienced hunger, true hunger. I think having these discussions sensitizes me and, hopefully, a lot of other people about what the issues are, how we can go forward, and how we can solve these problems. The question that I want to pose is, how can we move forward?"

In responding to Lemaux, the main participants were most articulate on what role they believed their adversaries should play.

Shiva: Options needed

"I think one very clear-cut criterion for going forward is to always posit alternatives at the time of any technology decision and any technology choice. So when a particular genetic-engineering option is being offered, to always look and ask is there another way and then allow farmers, consumers, and society at large to make its choices on the basis of real options being available. The reason there has needed to be the tremendous intensity of bringing up the alternatives that were excluded or made invisible was because biotechnology was offered as the only option for the future."

Prakash: Critics, give us your energy

"I don't think any responsible scientists, at least whom I have worked with, see biotechnology as the only solution, and we have always believed that this is an important tool in the whole range of things that we have. Let's accept first of all from the critics that this is a technology that doesn't bring any unique risks. This is a technology that compares with all the tools and techniques we have been using. Let's start examining where we could put our limited resources and start prioritizing it along with the other options that we have. . . . I think we can use the energy and the vision of the individuals that currently do not feel comfortable with the biotechnology. But we could indulge in a dialogue and learn from each of them and move on. I think all of us are very sincere here in this room in that we recognize the problem of hunger and are sincere in believing that there are solutions out there."

McGloughlin: No false barriers

"I absolutely believe that biotechnology is not the panacea to all the world's ills. We need to optimize all tools so that we can optimize the interaction of the various things that work best in a particular environment. However, we need to make this science-based. We cannot throw out the science. We cannot create false barriers based on pseudoscience or beliefs that are not compatible with using the best tools we have. We need to focus on science-based values that will allow us to reap the incredible capability and potential we have with all types of agriculture. Biotechnology is a very strong component of this agricultural tool case."

In October 1998, Michael Pollan offered unapologetic criticisms of biotechnology in a cover article in the New York Times Magazine *with the title "Fried, Mashed or Zapped with DNA." Pollan's piece was influential, helping to persuade foundations to contribute to antibiotech campaigns and, quite possibly, changing the tone of American coverage, as benchmark articles in the* New York Times *often have done. He was invited to speak at the hearing, and when given the floor, he sped to questions at the heart of the debate.*

Pollan: Don't lose sight of underlying politics

"I am thoroughly confused. I am not an expert in this field. I have a sort of different status than everybody here. I came to this subject as an amateur, as a gardener who wanted to plant one of these crops, which I did a couple of years ago. I grew some genetically modified potatoes in my otherwise organic garden, to explore what the implications were and what the implications of that experiment were for me. Uncertainty is a big theme.

"I would like to talk briefly about politics in this temple of politics, and that's something we haven't done very much. . . . Why are we discussing biotechnology and world hunger together? The answer to that question is political. This is an industry that is in a certain amount of trouble in this country and internationally. The problem I found in deciding whether I want to eat a biotech potato I had grown is, well, why should I? What are the advantages to me as a consumer? I could not find any good reason to eat this potato. It offered me nothing. It was a potato, a Bt potato. It offered, perhaps, the farmer something. It certainly offered Monsanto, the company that developed it, quite a bit. But the benefit to the consumer isn't there. The risks, the uncertainties, are there.

"When an industry is selling a technology to us and they cannot make a case to us of proven benefits, they have to come up with other arguments. The Third World hunger argument has been advanced by the industry. There is a suggestion out there that, by being critical of this technology, you are blocking the access of the Third World to something that may be useful to it. Let me assume it is useful to the Third World or is potentially useful. Do we as Americans have a moral obligation to the Third World that entails accepting this technology? I think that's a real question we have to deal with.

"Congressman Hall said at the beginning, that if this is about money I am not interested; if it's about feeding kids, I am. Unfortunately, those two issues cannot be separated. It is about both. We have to be very alert to the politics of this debate. Golden rice is the first crop that has come forward specifically designed to solve a Third World problem. It, too, is full of uncertainty; it's not ready to be commercialized. It's not even ready to be given away yet. But we are being asked to make our decisions as Americans with this in view. I think that's a question we all have to answer.

"I also want to resist the suggestion that is out there that being a critic of this technology, even being a consumer who does not want to use it, or someone who wants to label it, as Congressman Kucinich said, is, therefore, against it. Like everything in a democracy, science requires criticism."

At closing, David Beckmann summed up what he had heard. Biotech is another tool, he said, perhaps one that will prove useful. But sadly, both the technology and the debate seemed remote, in his view, from the needs of poor farmers in least developed countries.

Beckmann: A tool, not a solution

"One thing that I was struck by is that I think all the panelists agree that biotechnology is not the solution to world hunger. But I think they also agree that it is one possible tool that could be helpful in reducing hunger. They have very different assessments of the potential benefits and risks. They all agree, as Michael Pollan pointed out, that, in fact, there is a lot of uncertainty. And I think they agree there is massive neglect of other tools that could clearly help hungry people. . . . I am also struck that the biotechnology debate in connection to world hunger is taking place in a context of a tremendous imbalance of power and money, and that most of the biotechnology development so far—the research, marketing, even the controversy about it—has had nothing to do with hungry people.

"So far, the debate has been about the health and environmental concerns of people in the industrial countries, not the concerns of the poor woman farmer who is trying to feed her kids. At the end of the scale, that woman is tremendously vulnerable. How is she going to figure out what she should use so she just doesn't depend on other people to advise her? There's this massive imbalance of power and money, and it takes place in the context of equally massive public-policy disinterest in that woman."

Part Four

COMING TO GRIPS

19

COLOMBIA: RISING VOICES IN A TROUBLED LAND

Dazzled by the show of power, the common people did not discern the
covetous bustling which occurred on the rooftree of the house when
agreement was imposed on the town grandees. . . .
—Gabriel Garcia Marquez, "Big Mama's Funeral"

A STATUE OF Don Blas de Lezo stands at the entrance to the San Felipe fortress atop San Lazaro Hill, in the city of Cartagena. Blas de Lezo, a Spanish officer, had lost a leg in the Battle of Gibraltar. In Toulon, he lost an eye. Later the Battle of Barcelona claimed his right arm, but he fought on. Finally, in 1741, repelling a British invasion of 186 ships, he was wounded in his remaining leg—and what was left of Blas de Lezo died.

In 1999, 258 years later, the first global treaty proposed to regulate genetically modified food is suffering a similar fate near the spot where Blas de Lezo succumbed. One by one, vital parts of the Cartagena Protocol, the object of this United Nations–sponsored negotiation, are being hacked away: the labeling, the liability of manufacturers, the right to reject shipments—most of the powers and restrictions demanded by the buying world. Wielding the Big Machete at these talks is the United States of America.

Cartagena has long been a getaway point for pirates and plunderers, and that's how, three days into the talks, much of the world views the United States and its grain-exporting allies. From the start, Colombia, a spectacularly troubled country torn by revolution, land inequity, and cocaine commerce, seemed an improbable place to settle this dispute. And then, three weeks before leaders of the world gathered here, an earthquake of 6.3 on the Richter Scale had obliterated villages in Colombia's western coffee-growing region, leaving over one thousand people dead.

Yet there was a symmetry in meeting in Colombia. Why not come to the country with the world's richest storehouse of genetic resources to begin solving the problems of the Genetic Age? The nations of the world have harkened to the technology pounding at the door. Now, most of them have allied in the pursuit of regulations to govern the billions of dollars of gene-altered seeds and modified goods moving around the world. They have named this gathering grandly, the First Extraordinary Session of the Conference of the Parties, drawing authority from the Convention on Biological Diversity, an obscure, 1992 treaty so far spurned by ratifiers in the United States.

But the United States cannot get away scot-free; American-farmers and exporters would face restrictions from a protocol on shipping their grains to global markets. And biotech companies could have to accede to rigorous documentation when sending those bags of gene-altered seeds across international borders. So the Americans, too, have arrived at proceedings they had hoped to forestall.

The United States and Canada, its chief ally in this fight, argued that biological diversity and trade are unrelated. So what's to negotiate? But they were outnumbered, cornered perhaps. Now, what with the shift of fortunes in the biotech industry and the far-flung preparations for this gathering, organizers arrived with a reasonable expectation of compromise.

But as soon as representatives from 137 countries checked in, the wrangling began. The United States and its partners showed no intention of submitting to a trade-inhibiting document. The global politics seemed clear. Yet I am persuaded that in Colombia—a place where, literature has led me to believe, magic is possible—anything can happen.

In Colombia, a Greenpeace-led protest in the streets
of Cartagena that included a depiction of a tomato
engineered with fish genes did little to influence a United
Nations gathering of countries attempting to regulate the
flow of genetically modified products.

SOLUTIONS IN ORDER

This is a treaty the world desperately needs, I have concluded. The impasse
between the United States and the European Union seems unbreachable.
The skepticism that I witnessed in County Cork a year before hardened
first into resistance and then into a *de facto* moratorium on planting new
bioengineered crops. In Britain, the turn of events was stunning. The out-
of-power Tories had seized on genetic engineering and were bludgeoning

Prime Minister Tony Blair with it in return for the punishment their government had endured during the "Mad Cow" fiasco.

The British tabloids were spinning themselves into a Frankenfoods frenzy. By Monsanto's count, in January alone, a month before the Cartagena gathering, newspapers in Britain had published 940 stories on the conflict. Stan Greenberg, Monsanto's pollster, had been prescient four months earlier in his interpretation of a poll: "The latest survey shows an ongoing collapse of public support for biotechnology and genetically modified food," he had written.

The battle over GMOs was expanding into a global war. The multinationals' fears of a "Europeanization" of South America had proved well-founded. Brazil was refusing Monsanto's modified soybeans, having locked in the court injunction handed down while I was in São Paolo four months before. No genetically modified crops could be legally planted in Brazil—though farmers had been smuggling them in from Argentina, where Roundup Ready soybeans flourished. South American activists were outpacing their *Norteamericano* brothers and sisters. The ink wasn't yet dry on a Latin American Declaration on Transgenic Organisms signed in the high-altitude capital of Quito, Ecuador, by environmental, indigenous, and farm organizations. It began unambiguously: "We reject genetic manipulation as being an ethically questionable technology which violates the integrity of human life and the species that have existed on earth for millions of years."

In India, where Monsanto hoped its Feed the Hungry mantra would resonate, farmers had three months earlier torched fields of genetically modified cotton and anointed their attacks with a chillingly violent name: "Operation Cremation Monsanto." In Japan and Australia, calls for mandatory labeling grew louder.

Then there was the United States. American farmers already had placed many of their seed orders for planting around 71.8 million acres of genetically modified soybeans and corn in 1999—72 percent of the world's acreage of modified crops. "The year ends," I had written recently for my newspaper, "with a chasm between the attitudes of Americans and people elsewhere."

America—indeed, the world—could scarcely afford to let that chasm widen. Not with America's fifty-billion-dollar farm-export industry out on a limb. Not with anti-American sentiments deepening. Not with the technology crying for better understanding and regulation. Not with the survival of venerable institutions, among them the European Union, threatened.

Where were the leaders when they were needed? Vice President Al Gore, known for his environmental credentials, had jetted to Japan against the will of advisors in 1997 to rescue the Kyoto Protocol. But the United States had no interest in propelling a biosafety protocol. Jimmy Carter, an apostle for biotech, wouldn't be arriving in Cartagena. Nor would his former ambassador to the United Nations, Andrew Young, though he had envoyed for Monsanto on the issue. Rather than calling in the heavyweights to win a protocol, the industry had mobilized to weaken or kill it.

The United States and the biotech industry were relying on the World Trade Organization, not this uncontrollable alliance of nations, to sort out the deepening strife over GMOs. The WTO was the Big Stick they would use to force the world to accept genetically modified food. Yet these days, wherever I traveled outside the States, I'd seen hostility to the WTO. And there was talk in a variety of circles about voicing displeasure when the WTO gathered in Seattle nine months hence to plan the new round of negotiations. From what I had seen, the WTO lacked the respect to adjudicate any GMOs settlement.

For four years, environmental officials met at the compass points of the earth—Nassau, Jakarta, Geneva, Nairobi, Montreal—to set the stage for these negotiations. The goal was a set of far-reaching rules to govern trade in "living modified organisms," which has a different ring to it than what we hear in the United States—"biotechnology-enhanced food." Hardly anyone outside their little circle had paid much attention.

Until now.

LAND OF GARCIA MARQUEZ

I arrived three days into the negotiations and made my way to a convention center built of yellow volcanic stone on the Bahla de las Animas harbor. The negotiations were taking place a few blocks from the home of Colombian Nobel laureate Gabriel Garcia Marquez, whose novels of magical realism are full of unseen power, mindless bureaucracy, and absurdity. A good match, I reflected, as I waited for meetings that never happened and watched alliances sprouting like jungle vines. There were groups called the Friends of the Chair and Friends of the Minister, separately thrashing out the same issues and encountering one another only when they emerged from separate rooms for demitasses of thick Colombian coffee.

At one point, the scope had been so narrowed that a delegate claimed it could only properly be called the Cartagena Protocol on Animal Vaccines. What I saw was talk and confusion, not the signs of impending agreement: stare-down negotiations and urgent phone calls back to capitals.

But time remained. The United Nations had spent millions of dollars in pre-negotiating sessions on these talks, and the bureaucrats expected success. That's the way global governing works: Spend enough money and time in civil discourse, and things get solved.

"If we fail here," said Hamdallah Zedan, of Egypt, executive secretary of the Convention on Biological Diversity Secretariat, "we will not only damage the protocol on biosafety but undermine the convention as well." For another view, I asked Michael Williams, press secretary for the Geneva-based United Nations Environmental Program, what a failure would mean. With his characteristic bluntness, Williams, a New Yorker, replied: "It would be a major fuck-up after all that has been invested."

SPROUTING ALLIANCES

The negotiating world was divided into camps. The United States was part of the Miami Group, put together by the State Department's Rafe Pomerance, who had a title almost too long for a business card: Deputy Assistant to the Secretary of State, Bureau of Oceans and International Environment and Scientific Affairs. Pomerance was a past president of Friends of the Earth, an advocacy group that lined up with the planet's anti-GMO forces. If Pomerance harbored any such sentiments, he concealed them fully in Cartagena.

Preparing for South America, a strategy session between Pomerance and the Australians had expanded into a search for allies. Countries with the most to lose in Cartagena had a common bond: They exported grain. Pomerance considered Los Angeles as a meeting place for his fledging alliance, but he settled on Miami for its easier proximity. There, in the Florida sunshine in the spring of 1998, Argentina, Uruguay, and Chile joined the United States, Canada, and Australia as a new multilateral negotiating bloc henceforth known as the Miami Group.

The Miami Group's chief adversary in Cartagena was the European Union, whose representatives demanded a strict protocol. Central and Eastern European countries aligned in a separate negotiating alliance, and that left Africa, China, India, and the rest of the developing world, who had thrown in together, under the banner of the Like-Minded Group.

To clear the way in Africa, Monsanto had hired Andrew Young. There are few African-Americans who command more respect globally, and Monsanto, ever the well-connected company, had dispatched Young to Africa on two earlier occasions on company business. Young's team was "able to get our message out quickly," Monsanto's Robert Harness said over coffee in Cartagena.

Judging by the sentiments of Africans who came to Cartagena, Monsanto may have misjudged Young's impact. That was the assessment of Florence Wambugu, a Kenyan and a believer in biotech, who told me that she considered Young's journeys only marginally helpful. "In his own way, he created awareness. He talked to the right people, but he did not talk to the people who are here. I don't think he went far enough. I'm not sure he was aware of that," she told me.

The chief issues dividing nations were clear at the outset: The Miami Group, whose interests matched closely with those of biotech and farm industries, wanted a narrow scope that did not cover farm commodities or pharmaceuticals. They further demanded a so-called "savings clause" to prevent a protocol from interfering with WTO business.

To satisfy their clamorous constituents, the Europeans needed a toothy agreement, one that included a "precautionary principle" that enables countries to protect themselves even in the absence of scientific proof of harm. Developing countries, too, wanted wording enabling them to ban shipments, even on socioeconomic grounds. And they wanted respect, a commodity they considered scarce as the negotiations lurched forward.

I had noticed that people listened to an Ethiopian plant ecologist, Tewolde Berhan Gebre Egziabher. So I took him to lunch at an outdoor café on the day, it turned out, of his fifty-ninth birthday. I didn't know at the time that he would emerge as a key figure in the global food debate beyond Cartagena.

"We are afraid of becoming the dumping grounds for genetically modified organisms," he said, over a plate of beans and rice. "Our environment is much more vulnerable to certain interferences; that's because of our species diversity. Our life is more intricately linked with biodiversity than others' in the world. When it comes to genetic pollution, we are very vulnerable. We also have socioeconomic concerns: Local indigenous communities are going to be very easily disrupted. Say, for example, with the Terminator technology. The ability to manage seeds could completely stop." There it was again: the fear of being controlled and the fallout from Terminator, a public-relations disaster.

In the minds of most Americans, Ethiopia symbolizes famine. Yet Tewolde, as everyone calls him, rejected the promise of biotechnology to feed the hungry. "It will centralize production even further and only exacerbate the problem. We've already seen what happens with government interference in our thirty-year civil war. The research and development that could really help the poorest farmers is being ignored. I am not against biotechnology, but I do not believe that it can do the things they say."

Despite efforts made to persuade them otherwise, many of the Africans I talked with shared Tewolde's attitude. But there were conflicting sentiments: On one hand, they feared the environmental consequences of an unknown technology so heatedly debated in Europe. On the other, scientists in developing countries felt left out, convinced they were being denied a promising technology.

Finally, after five days in Colombia, I wrote a story saying that if an agreement were reached, it would exclude corn, soybeans, and the commodity crops, as the Miami Group insisted. In it, I quoted Rafe Pomerance on the prospects for reaching a protocol. "Maybe yes, maybe no," he told me. "The negotiations are at a very sensitive stage. There are many obstacles to overcome, plus the fatigue factor is setting in."

And there were just two days to go.

A BITING SNAKE

For Veit Koester, a Danish civil servant and the chairman of the proceedings, a biosafety treaty would be the capstone of his involvement in a decade of global environmental politics. Koester—a slight, earnest, pipe-smoking man who seldom smiled—had helped forge the Convention on Biological Diversity and had held chairmanships in other United Nations–sanctioned environmental efforts. He wanted an agreement. But he looked worried, and, as the hours ticked away, he was growing exceedingly frustrated with the United States delegation and what he saw as a commitment to block success. Perhaps in the world of environmental diplomacy, Koester had not experienced the hard-line negotiating tactics deployed when billions of dollars are on the table. Billions were indeed at stake in these negotiations, along with the freedom of multinationals to operate unencumbered by new global restraints.

Still, Koester expected an agreement because, almost always, people reach agreement in these forums. So he pushed and he gambled, presenting a chairman's text of unnegotiated provisions in hopes of forcing parties to compromise. He miscalculated, misjudging both the resolve of the principal combatants and the alienation of developing countries.

The State Department's Melinda Kimble, who headed the United States delegation, showed me how deeply the Miami Group had dug in when I sat in one night as she met with environmental advocates. "I want to make very clear that it is the position of the United States government that we do not believe there is a difference between GMO commodities and non-GMO commodities. We do not believe that you should have discriminating trade just because you do not want something to come into your country. We think free trade makes everybody better off," Kimble said, summarizing the United States position.

Kimble was a study in inflexibility, laying out many whats but few whys. She and Rafe Pomerance were able negotiators; but their unflinching opposition frustrated and antagonized delegates from every corner of the globe. Koester complained to me that the American negotiators were impolite, even rude.

Pomerance later disputed Koester's assessment while allowing, "Sometimes people get stressed with no sleep. There's a lot at stake."

With time growing short, Greenpeace, which had sent ten representatives to South America, turned up the heat. At a demonstration outside the convention center, protesters dressed as the genetic mutants Fishberry, Rat-Trout, and Scorpion-Maize and carried placards that read, *No a la contaminacion genetica*. A Greenpeace news release accused the United States of seeking a biotrade agreement rather than a biosafety protocol. "It seems that the U.S. is willing, in cold blood, to threaten biodiversity in the name of short-term profit interest," it read.

Things were getting tense and uncomfortable. Andy Pollack, the *New York Times* correspondent, and I were asked to leave a European Union reception one evening. Parts of the convention center suddenly were blocked off by plainclothes security after being open for days. People were dropping from gastrointestinal illness. Monsanto's delegation was especially vulnerable, and the company's Christie Chavez was taken to a hospital. Each time Koester would announce a plenary session to take stock of the proceedings, it would be canceled.

With odds against success, Koester finally convened a plenary. But rather than stimulating consensus, his "chairman's text" triggered an outpouring of recriminations. One by one, official representatives from more than fifty countries stood to denounce it. A delegate from Mauritius called it a joke. A Guatemalan said "My delegation is shocked at the way things have been going. We have been insulted in terms of our intelligence."

Togo's environmental minister, Koffi Dantsey, said that the chairman's text "represents a snake that is about to swallow you up. When you get close to a certain snake, it will bite you, and you will die," he said.

Veit Koester was stung. "I think the general trend is rather clear now," he said from the dais. "But do not come and tell me that this document is not the result of a democratic process. Do not do that; I do not accept it. It's up to you to negotiate, not to me. If you do not want to negotiate, how can I force you? You have to know that this situation would come sooner or later. You have to try to find a balance, sooner or later. Be reassured, I like you all very, very much. All of you, all of you. You should not be sad. I am not sad. You should not be embarrassed. I am not embarrassed. You should be proud of what you achieved."

I filed a story saying that hopes for a treaty had diminished. "The protocol is very ill in the hospital, but perhaps it can still recover," I quoted Pomerance as saying.

That night, there was one last opportunity to succeed, and Koester admonished the delegates to try. "Stop quarreling. Stop arguing and make the protocol," he said.

HUNTING FOR MAGIC

In sessions that ran past midnight, most of the world did reach an agreement, which ordered labeling and a new regimen for world trade in modified products. But without the Miami Group behind it, the agreement had not a prayer.

In truth, nothing had been negotiated, and what occurred for thirteen days in the yellow convention center and in the Santa Clara Hotel, where the North Americans set up their operation, could only be called discussions.

"No deal in this case was much better than a bad deal," Pomerance declared in the early-morning hours.

Long afterward, the excitable Pomerance, who had left the State Department, hadn't changed his tune. "There were provisions in the goddamn protocol that weren't adequately negotiated. Not because people
would disagree but because the thing wasn't managed properly in terms
of focusing people on issues. The chairman just kept us sitting there, it's
as simple as that. The only goddamn reason it came to an end was because
we had planes to catch. There was just time being used up to see if anyone
would break. It was just very frustrating because we didn't need to be
there. We were all exhausted."

With collapse all but assured, hospitality in the volcanic stone hall had
eroded along with hope. For many hours there had been no food nor
coffee, and now drinking water was running out. The last bite many
people had eaten was a square of white cake that Beth Burrows, an anti-
GMO activist from the Edmonds Institute in Seattle, had brought to the
gathering as a peace offering. Michael Williams, the U.N. press secretary
assigned to the affair, produced a bottle of rum, which several of us poured
into coffee cups. All that remained were recriminations, which I recorded,
wearing a rum buzz, near dawn.

Joseph Gopo, a scientist in a gray, Italian-cut suit who headed Zimbabwe's biotechnology office, accused the Miami group of "cruel" tactics
and of imposing its will on the world. "What we're being asked to do is
sacrifice internationalism to the point of view of a few countries," he said.

Tewolde, the Ethiopian, predicted that the United States refusal to
budge "will make developing countries more and more suspicious of genetic engineering."

Veit Koester's chance to be recalled as the man who forged the first
global agreement on modified food had disappeared, and he looked crestfallen. I asked him to explain what had gone wrong. He lit his pipe before
answering. "There was a lot of money involved; huge amounts of grain
and food moving around the world," he said.

Amid the wrangling, two seeds were planted in Cartagena that could
sprout another day. The organizers kept the session alive, promising to
reconvene. It sounded to me like face-saving, and I reported that "many
in Cartagena wondered if they could regain the momentum that brought
them to this Caribbean seacoast city." The second seed was the selection

of a new chairman, a Colombian named Juan Mayr Maldonado, who was his country's environmental minister. He was a man known both for his affability and his battle-tested skills in squeezing compromises among revolutionaries, drug-traffickers, and landless people. Juan Mayr was accustomed to dealing with sensibilities in the extreme, which—given the deepening chasm dividing nations on genetically modified food—seemed the appropriate résumé.

Rather than narrowing the world's differences, the gathering in Cartagena illuminated them. To many, the United States emerged a villain. Andy Pollack began his dispatch for the *New York Times,* "Attempts to forge the world's first global treaty to regulate trade in genetically modified products failed this morning when the United States and five other big agricultural exporters rejected a proposal that had the support of the rest of the roughly 130 nations taking part in the talks."

The story I filed after two hours of sleep led with these words: "U.N.-sponsored negotiations to write rules for genetically modified organisms collapsed Wednesday amid deep divisions over the powerful new technology."

TRADE VERSUS THE ENVIRONMENT

The breakdown, I began to understand, was more than dissent over gene-splicing. It demonstrated more clearly than ever the new competition in the world between the forces of free trade and environmentalism. The passionate discussions that bore no fruit in Cartagena had amounted to an initial sorting of these global imperatives.

At the Earth Summit in Rio de Janeiro in 1992, governments agreed to attack environmental ills around the globe and later produced the Montreal Protocol to phase out air pollutants. But agreement is more difficult in the new economic order of international commerce.

Pomerance observed that the Cartagena talks held "profound consequences for the World Trade Organization and the world trade regime. I heard for myself from a number of delegates walking around the room tonight," he said, "that they in fact wanted this agreement in order to avoid having to base their decisions in a way that would have to be sustained under the World Trade Organization."

The collapse in Cartagena also raised questions about whether the United Nations can provide a forum for dealing with powerful genetic technologies. A separate United Nations negotiation, a Food and Agriculture Organization initiative on genetic resources—called an undertaking—looked in need of an undertaker after collapsing the year before in Rome.

The Miami Group had prevailed in Cartagena. But without an international agreement and no prospect of one, grain-exporting nations and the biotech industry departed Colombia with an even more difficult burden of deflating global skepticism. "We have to do everything we can to inspire familiarity and trust and then do nothing to undermine that trust," Willy de Greef, director of regulatory affairs for Novartis, told me before catching his plane.

Just about every weary delegate I tracked down sounded pessimistic about righting the negotiations. Except for one—Juan Mayr, the new chairman. "It has been a great advance," he said, drawing from a mysterious storehouse of optimism. He said that he planned to begin seeking "a special mechanism" to resolve the issues dividing countries. "In some ways, right now there is a balance," he added.

I didn't understand what he meant, but after thinking about the exigencies of the day, as well as about Mayr as the new leader, I contradicted myself in a second-day analysis. "When trade issues collide with environmental concerns, who wins?" I asked in the story. "In Cartagena trade carried the day. But with global concerns about genetic engineering mushrooming, the outcome of the next negotiation could be different."

Much would have to happen first. By the time I flew back to Washington, I had concluded that the road to a treaty on genetically modified food twisted through Seattle, where the World Trade Organization was laying plans for a gathering that would be memorable indeed.

PLANTINGS TEN ◁▭▭▭▷

I heard that the Canadian federal court had ruled in Percy Schmeiser's seed piracy case, so I telephoned him. He had turned seventy, and much had happened since Monsanto accused him of sprouting Roundup Ready canola without paying the company its technology fees.

Percy had become an international celebrity. Farmers around the world were tuned in to his plight, and he had told his story in twenty countries, traveling as far as India and Australia at the behest of farmer associations and anti-GMO organizers. In his audiences, people would shake their heads when he got to the part about private detectives coming onto his land. They would groan when he told them how, if a company determines they have a patent on what you're growing, your crop might belong to them.

Percy had been a politician and a mountain climber. He was resilient in ways many people aren't. But the travel, the notoriety, and the uncertainty weighed heavily on him, I had heard. He'd sent his wife, Louise, to California to get her away from the hubbub after she'd developed high blood pressure and had a fall. When I called, he had been preparing to head to South Africa for another round of speech-making. But after what the judge had ruled eighteen hours earlier, Percy said he probably wouldn't be traveling anywhere right away. He had some thinking to do.

He had lost, and Monsanto had won. The federal judge in Saskatchewan ordered him to pay Monsanto about $15,000 in damages and an amount to be negotiated as reimbursement to the company for profits on his 1998 crop. When I tracked down the court ruling, I noticed that the judge had not ruled directly on Percy's contention that the crop had sprouted from seeds blown onto his land from trucks or as a result of pollen from neighbors' crops. In passing, the judge had cast doubts on what Percy had maintained. But according to the ruling, the

legal question was whether Percy had known his crop originated from Monsanto's patented Roundup Ready variety. And
no matter the source of the seeds, Percy ought to have known,
the judge wrote.

The money at stake was less important than the message the
case sent. Around the world, farmers were outraged by biotech
farming's prohibitions on saving seed for future crops. The technology fees rankled them. In Argentina, soybean growers simply had refused to pay them, triggering a robust black market.

When I telephoned Val Giddings at the Biotechnology Industry Organization, he said he regarded the decision as significant for patent protection. "It's clearly an important ruling
for the industry in North America with potential implications
and ripple effects throughout the world," he remarked.

To the industry's dismay, the message that had gone out in
press accounts was something other than the sanctity of patents.
A *Washington Post* story asserted that the canola had gotten on
Percy's land "apparently after pollen from modified plants had
blown onto his property from nearby farms." Several farm
leaders I talked to were shocked, worrying aloud that Percy's
fate might befall unwitting corn farmers in the United States.

It was early in the morning when I reached Percy at his farm,
and he had slept little. "It's a terrible position that this case is putting farmers into, just unreal. I guess it's just tough luck if you get
Monsanto seeds on your land," he said. He told me that he had
spent $200,000 fighting the charges and that he hadn't decided
whether to appeal. "The law isn't made for individuals and regular people. I don't see how any person, any farmer can stand up
to a multinational company in court," he said.

Percy was beaten but unbowed, and I suspected from his
words that the industry could pay a price in the future for their
court victory. "Monsanto might have won this battle," he said,
"but I don't think they've won the war."

20

THE BATTLE OF SEATTLE

*Fear of globalization is really a concern about who makes the big
decisions, and who suffers from the big decisions.*
—Todd Gitlin, Sociologist

*GMOs have brought together the bean-sprouters and
the snuff-dippers.*
—Jim Hightower
Former Texas Agriculture Commissioner

AFTER OUR "LOCKDOWN" at Pier 66, we snaked through downtown Seattle in a Range Rover, diverted again and again. When I had arrived at a glass palace along Puget Sound for the forum on genetically modified food, police had kept us standing outside in the rain for fifteen minutes. When it ended, the police kept us waiting inside until a commotion ebbed in streets nearby.

Once we were moving, police in visored helmets and Robocop armor blocked us from motoring anywhere near the Seattle Convention Center, venue of the World Trade Organization's ministerial hearings and, at this moment, the epicenter of a worldwide antiglobalization movement. It didn't matter that a United States congressman rode shotgun in our vehicle.

The congressman, Dennis Kucinich, a Democrat from Ohio and an irritant nonpareil in Washington to the biotechnology industry, had joked to his audience about the state of emergency in the city. "Those of you who have spent time in federal facilities will be familiar with the term 'lockdown,' " he deadpanned. The loss of the basic right of movement in a twenty-five-block "security zone" and a 7 P.M. to 7:30 A.M. curfew were among restrictions invoked after protests had engulfed the gathering of trade ministers.

The WTO and United States officials had known that protesters were coming from around the world: The Ruckus Society, which is based in Berkeley, California, had publicized its training sessions; left-leaning magazines had trumpeted the approaching "Battle of Seattle." In my office in the other Washington, three thousand miles away, I had received daily faxes reporting new fleets of buses chartered for protesters and more caravans of cars and trucks—one sixty vehicles long—mobilizing to depart. One missive informed me of thirty angry metalworkers en route from France.

In Seattle, I remembered the words of a young Muhamed Ali, pontificating on the absence of stealth in George Foreman's powerful punch. "George telegraphs his punches. Look out; here comes the left. Whomp! Look out; here comes the right. Whomp! Get ready, here comes another left," the wide-eyed Ali told reporters with mock fear before the two heavyweights met in the ring. WTO organizers knew the punch was coming. Unlike Ali, the WTO and the city of Seattle couldn't get out of the way.

Kucinich, like Ali, had been a *wunderkind* in his trade before enduring crunching setbacks. His political career had rebounded since those ignominious days in the 1970s when, as the "boy mayor" of Cleveland, his city had plunged into financial default. He was in his fifties now, but he looked younger in his pageboy haircut—and he still had the reputation of a hell-raiser. To the establishment, he was the worst kind of hell-raiser, one rooted in a safe seat, as he demonstrated convincingly in the most recent election, winning two-thirds of the vote over a Republican salesman named Slovenac. I'd always enjoyed his humor back in Washington, D.C.; at an orientation for first-term House members, he handed out trading cards with his photo as a four-foot-nine, ninety-seven-pound third-string quarterback. Another time, at a banquet, the previous speaker, the late Republican congressman Sonny Bono, had told him to break a leg. Kucinich responded, "Look, my dad was a Teamster. We don't kid about that."

In his speech in Seattle, Kucinich wasn't kidding when he warned the biotechnology industry not to fight his newly filed legislation to put labels on packages of food that contained genetically modified ingredients. "If they resist strongly enough, it's quite possible that we'll just pass up the issue of labeling and go right to a ban," he said from the dais. He was somebody the industry needed to keep an eye on, and in the audience, a Monsanto government-relations specialist, dressed in running shoes and windbreaker, wrote Kucinich's words in a spiral notebook.

What Kucinich wanted in the immediate future, however, was reasonable proximity to the front door of the Westin Hotel, where he was staying, and respite from this mayhem in the streets. He also wanted to avoid a broken leg which, despite the congressional pin he wore in his lapel, was a possibility in such mayhem. As we headed toward his hotel, we were diverted anew on our approach to Seattle's famous Pike Place Market, the latest arena where police and demonstrators clashed. Clouds of tear gas trapped low to the ground by moisture wafted toward us as police yelled at us to stop.

In forty-eight hours, the Seattle police force had evolved from a permissive lot, refusing to arrest protesters for civil disobedience, into a menacingly combative force augmented by the National Guard and firing gas, wooden-block projectiles, and rubber pellets. They dressed in all black from their helmets with darkened face shields down to their reinforced shoes. They wore shoulder pads, elbow guards, and padded protective vests along with shin and knee guards with reinforced articulations. From their waists hung extra-long riot sticks and plastic wrist-ties for rapid subduing.

In addition to their 9mm pistols, some of the officers carried tear-gas launchers that looked like space-age Gatling guns. They had two methods of delivering pepper spray: handheld devices for close-range applications to the face and air guns powered by carbon-dioxide cartridges for distance shooting. By now, many city police, SWAT teams, and state police were in vile moods from long hours and endless provocation. It was wise to avoid them; it didn't matter who you were: A United States senator from Illinois, Dick Durbin, had been prevented by police from entering Kucinich's hotel and then admonished for taking his morning jog. When we talked that morning, Durbin, whom I've known for twenty-five years, used the words "armed camp" to describe the Seattle streets. "I never thought I'd see a scene like this with military troops in an American city," he told me over the phone.

Transporting Kucinich to the Westin was a logistical feat. But I insisted it could be done because I had negotiated a similar route the night before to deliver a scientist to her hotel after a probiotech gathering. It was my first occasion to be choked and blinded by tear gas.

I had thought that I was being clever scooting through alleys to transport the scientist, Irish-born Martina McGloughlin, of the University of California at Davis, who was recovering from a broken foot and carting around a suitcase that seemed to have been packed with textbooks. I'd gotten her to the front door, and I was proud of it. But traveling the same route out of the downtown, I ran into a crew of black-clad protesters who had overturned a Dumpster and were being chased in an alley by fourteen or so riot police, firing gas. It looked like fog and smelled like fireworks coming through the vents. I rolled down the window, which was my first mistake, and when the chemical fog poured in, I couldn't breathe.

My second mistake was to hit the brakes when, in my rearview mirror, I saw that one of the cops had stopped running after them and was coming back toward me. Good, I thought, he would direct me out of here. Instead, he banged on the trunk of my rented car, a Dodge Intrepid, swinging his riot stick like a man chopping wood. I jammed down the accelerator and fish-tailed away, veering so wildly when the alley opened into the street that I bounced over a curb. My tire went flat a few blocks later, just outside the security zone. I pulled the car to the curb and left it.

I was glad I wasn't driving now, at the intersection of Seneca and Eighth Avenue, where a half dozen officers directed traffic away from downtown. A white sedan directly in front of us refused to be diverted. The driver, an African-American, thrust something toward the police, perhaps a business card or an ID of some sort. One of the officers moved toward him and another gestured angrily toward us to move. "Wait," Kucinich said to the driver of the Range Rover, Francesca Lyman, of MSNBC. "Let's see what they do to him."

The car was a boxy Oldsmobile with its only noteworthy element a bumper strip in the left rear of the back window reading MCIVER. The driver insisted on turning left, toward the downtown, but the police were unpersuaded. Voices grew louder. Then a cop who had rushed forward from the sidelines yanked the man from the car. Another officer strode toward us, yelling "GET GOING." Francesca wheeled away. "I'll bet you that guy either is an elected official or ran for something," the congressman remarked.

The next day, my friend Kathleen Best, who was metropolitan editor of the *Seattle Post-Intelligencer,* described to me over the phone her budget of stories for the next day, a day when a six-column headline on page one would proclaim: "Crackdown—And New Clashes."

"And you won't believe this one." Kathy laughed. "The Seattle cops roughed up the only black member of the Seattle City Council."

I thought about an Illinois state senator I had known who, while visiting Cambridge, Massachusetts, suffered a heart attack. The man walking just behind him was a cardiologist, and the second passerby was a physician. The senator survived.

Seattle City Councilman Richard McIver enjoyed similar luck. As witnesses for his allegation of police brutality, he would have not only a United States congressman but also two journalists, all of whom later would be called upon to give statements to the Internal Affairs unit of the Seattle Police Department. By then the police chief, Norm Stamper, had resigned under attack from the public, protesters, and fellow police. What's more, McIver's passenger was his lawyer. Along with Kucinich, I ended up in the *Post-Intelligencer* article, observing that "one of the officers pulled him out of the car, and not gingerly."

McIver had the memorable quote. "I'm fifty-eight years old. I had on a four-hundred-dollar suit," he told the reporter. "But last night, I was just another nigger."

A "STAR-CROSSED" AFFAIR

By then, the World Trade Organization gathering seemed close to collapse. Trade ministers from 135 countries had gathered to set an agenda for the coming round of global trade talks, to be called the Millennium Round. But despite the warnings, the WTO, the Seattle Police, and, indeed, the world had been caught unprepared. Upwards of forty-thousand protesters clogged the streets, among them a phalanx of masked, window-smashing, and graffiti-spraying anarchists. The protests, as best I could tell, were sapping the WTO's political will.

The Tuesday morning opening of the ministerial meeting had been delayed. By 7:30 A.M., protesters had chained themselves to concrete barriers in front of the convention center and blocked an off-ramp from Interstate 5 to downtown. By eight o'clock, protective fences had been destroyed, and by nine, confrontations between police and protesters played out in a fog of acrid gas. At 9:50 with the opening of the talks on the verge of postponement, a reporter asked Chief Stamper if the protesters had won. "I don't view this as a game," he replied. "And I don't know that this is a win-lose situation." That was before twenty thousand prolabor protesters marched from Memorial Stadium to downtown.

It was a miserable week for just about everybody, including five thousand journalists, with street explosions, impossible travel, and botched global business, often as a cold, sideways rain assaulted the city. Besides the fog was an air of foreboding and meanness that seemed to numb the senses of analysts, mine included.

But I believe that I finally understood. What happened, I am convinced, was a seminal event in global politics. In the alliances that formed around the WTO talks, between labor and environmentalists, between consumers and farmers, among activists from different countries, a new global populism took form. What the world saw just weeks before the new millennium was the emergence of a new force that promised to exert itself not only in trade but throughout multinational governance. Resistance to genetically modified food, the world saw in Seattle, was a unifier in this new world politics.

The scene in Seattle was a far cry from the opening of the previous round of trade negotiations, in 1986, in the quiet seaside town of Punta Del Esta, Uruguay. No protesters or television cameras showed up. During the Uruguayan Round of trade negotiations, few Americans without a direct stake in the proceedings followed the workings of the Geneva-based WTO. I doubt many people even knew what WTO stood for until the Seattle gathering served up America's most chaotic street scenes in a generation.

The WTO was created in 1994 among the parties to the original General Agreement on Trade and Tariffs, or GATT, and bestowed with something the GATT lacked: the capacity to enforce its rules in international commerce. In more ways than one, the WTO is a government club. If a country violates the rules with unfair trade practices, it will end up paying the complaining nation—as Europe did in the late 1990s, in punitive tariffs, for refusing to allow imports of U.S. hormone-fed beef.

That case was the first significant food-safety dispute under the 1994 WTO agreement on the application of Sanitary and Phytosanitary measures, known as the SPS Agreement. In 1995, the United States contended that the European Union's ban on hormone-fed beef from the United States violated the SPS Agreement because it was not rooted in scientific evidence and therefore arbitrary. Eventually, the WTO Appellate Body agreed with the United States that the ban was a trade barrier, but Europe ignored the ruling. In May 1999, the United States retaliated by imposing tariffs of $116.8 million against several European products, among them foie gras, Roquefort cheese, and Dijon mustard.

The WTO is, if you believe its detractors, an elitist organization operating at the behest of multinational corporations and a threat to workers and the environment. But it also is the sole court for settling international trade disputes. And the WTO is, most certainly, the future battleground for the world's colliding policies on genetically modified food. As a United States trade official put it, global confrontations over genetic engineering will make the United States-Europe ruckuses over bananas and beef during the end of the 1990s "look like peanuts."

In Seattle, the mission of the trade ministers was not to issue rulings about genetically modified food or to adjudicate complaints about Europe's biotech policies. Like a choir arranging the sheet music to sing by, the ministers had convened merely to set an agenda for the next three years of negotiations. But the issue of biotechnology lurked near the surface. With the collapse of the drive for European consumer acceptance, the stymied companies, along with an equally frustrated United States government, were peering down a tunnel with no light visible.

The biotechnology and food industries needed a victory, and the gathering of trade ministers opened the door to some adroit maneuvering. In less than two months, the next and possibly last effort to forge a global biosafety treaty under the auspices of the United Nations would commence in Montreal. After the breakdown in Cartagena, the United States, Canada, and their allies in the Miami Group had pulled farther from the rest of the world, which was demanding caution—and labeling—of genetically modified foods. Those anti-GMO feelings could well prevail in Montreal, the industries understood, forcing them to endure a stringent global treaty. What some in the industry told me they wanted in Seattle was a way to undermine the Montreal proceedings.

That device, proposed by the United States and Canada, was a special WTO working group set up to analyze how WTO policies affect modified food and, ostensibly, head off trade disputes. It had the potential of real value if it could hasten an end to Europe's *de facto* moratorium on new approvals of gene-altered inputs. But a working group had a more immediate benefit: giving the Miami Group cover in Montreal for further intransigence. The United States and its allies could rightly ask: Why do we need a biosafety treaty when we just agreed to set up another global panel in Seattle to deal with these issues?

At the Seattle Convention Center, when delegates finally got inside, most issues on the table seemed too weighty or too parochial to be resolved. The United States had pushed a pro-worker initiative that was clearly out of step with global sentiments. Likewise, the United States continued to demand that Europe slash the protections for farmers in its Common Agriculture Policy. Meanwhile, India and developing nations wanted to take up old grievances from the Uruguay Round before considering further trade liberalization.

But movement on biotechnology seemed possible. And it was stunning news, barely reported amid coverage of the street clashes, when the European trade ministers signaled that they might agree to the working group on genetically modified food. For two years, as activists and European consumers built their anti-GMO movement, I'd watched European governments build a wall to keep out modified food. For Europe to accede would be a surprising development, quite possibly the sole tangible achievement in this star-crossed gathering.

PROBIOTECH BLITZ

Meanwhile, in hotels and meeting halls around town, opponents in the debate fought the battle of public opinion. The Seattle meeting will be remembered for civil unrest. But it will also be recalled as the moment when, after months of preparation, supporters of biotechnology made their stand. Until now, a combination of ignorance, arrogance, and jealousy of the competition had prevented the industry from fighting back effectively. But in Seattle, the biotech-food alliance drew support from scientists, agribusiness interests, members of Congress, and even President Bill Clinton to wage a potent public-relations campaign.

Clinton was shaken by a separate tempest of his own making when he told another of Kathleen Best's *Post-Intelligencer* reporters in an airplane interview that workers' rights ought to be enforced by as-yet-nonexistent WTO rules. The president had sprinted beyond the position of American negotiators in a concession to organized labor possibly intended to gird his vice president, Al Gore, for the coming round of presidential primaries. The president triggered the ire of many delegates, especially from India, Brazil, and Egypt, who were not yet persuaded that worker conditions should be studied. Amid his incendiary words about labor and his own criticisms of the WTO structure as closed and inaccessible, Clinton took time to deliver the highest level defense to date of genetically modified food.

In his speech, the president promised to "never knowingly permit a single pound of any American food products to leave the country if I had a shred of evidence that it was unsafe, and neither would any farmer in the United States of America." Clinton added: "I say to people around the world: We eat this food, too, and we eat more of it than you do. Now, if there's something wrong with anything we do, we want to know about it first. But we need to handle this in an honest and open way."

The biotech forces were thrilled. Unfortunately for them, the president's message drew little attention from a global press corps consumed by in-the-street explosions. In normal times, an endorsement from the president of the United States would have been a public-relations coup. But the wafting gas, smashed windows, and combat between protesters and police offered stunning visuals that had relegated the substance of the talks to a sideshow.

Meanwhile, food retailers had launched the so-called Alliance for Better Foods, pumped up with a million-dollar-plus budget. They showed up in Seattle in full force. Industry representatives met each morning at the Madison Hotel in yet another alliance, bringing in members of Congress and scientists to press their case to reporters. None of the biotech advocates came more swiftly to the point than Congressman Charles Stenholm, D-Texas. "Empty your pockets," he said to breakfasting industry representatives, admonishing them to fight the critics of genetically modified food. "We have to take to the streets ourselves. Otherwise, we lose," he said.

A senator who goes by the name of Kit, Christopher S. Bond of Missouri, assembled a forum of scientists. In front of them, he unveiled a placard with the names of three hundred scientists who had signed a letter to trade ministers decrying the opponents of modified food. I recognized at least twenty of the names as employees of Monsanto. Scientists in his retinue, among them plant biochemist Douglass Randall of the University of Missouri, claimed a new resolve to join the public debate. "We're very good at communicating with one another. Now we have to step up," he said. But only a handful of reporters showed up for the probiotech show.

Bond, who had tasted the gas himself on the way to a meeting, lashed out uncharacteristically at companies not signed up for the fight. In a speech that belied his usual devotion to Chamber of Commerce unity, he singled out Archer Daniels Midland, the Decatur, Illinois–based agribusiness giant, which had recently sent tremors through the biotechnology industry by offering to pay farmers a premium to grow conventional crops rather than genetically engineered varieties.

"What we are concerned about is that some American companies are taking a dive," Bond said.

AFFINITY OF PROTESTERS

Gene-altered food was a key issue at WTO but just one of the issues sending people into the streets. About half of the forty-thousand protesters, representing 140 countries—more than the trade ministers—had arrived at the behest of labor unions; the other half featured an amalgam of environmental groups, consumer advocates, and peace groups, sprinkled with such militants as the Black Army Faction of Eugene, Oregon, whose members tagged Seattle businesses with the anarchist symbol, the encircled "A." Protesters moved through the city like huge, shifting amoebas, blaming the WTO for a host of ills: disappearing rain forests, shrinking biodiversity, and refusing to protect wildlife, notably sea turtles, threatened anew by a WTO ruling undercutting the United States' order that shrimpers use Turtle Excluder Devices, called TEDS. Police had their black-padded suits and their modern tools, but the protesters successfully deployed high-tech tools of their own: the Internet and cell phones.

Alliances formed for the new millennium. Labor and environmental groups stood together amid wafting tear gas. Middle-aged Teamsters clad in black-satin jackets emblazoned with the union logos and the numbers of their locals walked alongside twentysomethings who had strapped on green turtle backs or had climbed into black-and-yellow monarch butterfly suits, a reminder of the Cornell University study discovering threats to monarch larvae from modified corn. Gay rights activists and anti–death penalty campaigners marched with advocates of forgiving developing countries their debt. In a reemergence of affinity groups, protesters driven by separate causes joined forces to magnify their impact, creating a model that would return to political conventions the next year in Philadelphia and Los Angeles.

In Seattle, the place of genetically modified food in the modern environmental movement crystallized. Unlike their European counterparts, mainstream environmental groups—the Big Ten, as they are sometimes known—had all but ignored the issue until 1999. In the early years of the development of modified plants, some of the dominant groups bought into the industry's promise that genetically modified plants would cut the use of farm chemicals. Others lacked the money and the experts, and therefore the credibility, to commit to an issue that stretched beyond the environment into farm policy, trade, and health. In the mid-1990s, with the gene-altered seeds sprouting, advocacy groups in the United States were pinned down in a rearguard action trying to stave off a rollback of earlier environmental protections, from clean air to endangered species, threatened by the Republican-held Congress.

The anti-GM movement that coalesced in Seattle had separate components: environmental advocates challenging an unproved technology; left-leaning trade groups condemning the patenting of genetic technologies as exploiting the world's poor; consumer advocates pressing demands for mandatory labeling of genetically modified foods; and farmers concerned about multinationals controlling what they grow.

Bill Christison, who cultivates two thousand acres of corn and soybeans in western Missouri, stood in Seattle alongside Jose Bové, the French farmer who had orchestrated an attack on the McDonald's in Cavaillon, France, the summer before. I recall the photo in the Sunday *New York Times* after Bové's charge: Ronald McDonald lay felled outside the franchise in a mound of playhouse plastic balls. Bové was a celebrity now, trailed by a film crew; during his coming trial in France, where he was convicted of the attack, his acolytes would deliver him to court each day in an oxcart amid cheers from his supporters. In Seattle, he arrived at the McDonald's with several hundred pounds of Roquefort cheese that he had brought from France without paying the extra tariffs ordered by the WTO as punishment for Europe's refusal to accept hormone-fed beef from the United States. When I walked by later I saw a window had been smashed and on the plywood patching the hole someone had spray-painted the words "McShit Meat is Murder."

In Seattle, Christison and members of his National Family Farm Coalition spent time "building relationships and solidarity," as he put it, with members of *Via Campesina,* an international organization of farmers. What brought them together was fear of losing control as a result of the multinationals' worldwide acquisition of seed companies. Already, farmers in the U.S. and Canada wanting seeds modified for herbicide tolerance and insect protection were obliged to pay "technology fees" and sign contracts promising not to reuse these seeds. In Christison's eyes, these arrangements hastened the day when grain farmers in the U.S. sink to the status of "hog-house janitors" in corporate-run livestock operations.

"We are going to be controlled by producers like in the poultry industry and the factory farms, and we will all be beholden to some corporation," Christison told me in Seattle at an old Christian Scientist church rented by advocacy groups as a headquarters. "When family farms are gone, people will pay a pretty penny for their food. And it's not the kind of food that they want to put in their bellies."

Meanwhile, Jim Hightower, the former Texas agriculture commissioner and talk radio's voice from the left, exhorted hundreds of farmers. "We're here to tell Monsanto and the other corporations: Get your modified seeds and your chemicals out of our food supply."

The world's anti-GMO stalwarts came to Seattle. Vandana Shiva, the Indian physicist, author of *Biopiracy* and other antibiotech books, warned of intellectual property rights being gathered up around the world by life-science companies. "First they patent it, saying it is unique. But then they argue that it's identical, or substantially equivalent," Shiva said, mocking the insistence by the biotechnology industry and the United States government that modified foods need not be so labeled.

"We need a five-year moratorium not just to prove the safety of GMOs but to let these people get their heads together," Shiva added.

A POLITICS OF CULTURE

Combatants who appeared alongside Shiva in one of Seattle's antibiotech sessions testified to the global makeup of a new movement: Tewolde Egziabher, general manager of Ethiopia's Environmental Protection Authority and a leading force in the United Nations biosafety talks; Mae-Wan Ho, a radical scientist from the Open University in the United Kingdom; David Bryer, who directs the United Kingdom office for Oxfam, the relief agency; and Congressman Kucinich, the leading anti-GMO voice in Congress.

Later, some of the participants showed up at the Dahlia Lounge for a respite from the convention and a GMO-free dinner. The chef, Matt Costello, said he had prepared the meal to avoid any known ingredients that came from genetic engineering. "It's really important for you to be able to know what's in the food you eat and how it's made," he said. Then he laid out a spread that included Alaskan spotted prawns on a bed of locally grown, Washington State pea pods; pumpkin tortellini made from organic flour; and wild salmon.

I asked Kucinich why he'd leaped into an issue of such complexity, a matter that is certain to put him crossways with the farm lobby in his Midwestern state. "I see this as a transcendent issue," he replied. "This really is more important than just food." I told Kucinich I didn't fully understand what he meant by that, and he explained his thinking:

"The food we consume describes who we are: our ethnicity, our culture; our dietary preferences; our religion. Food is not simply a fuel. We truly are what we eat. And what we eat determines what we're going to become, and that's true not only in terms of individuals, and their health, but it's true for the human race.

"We're in an era now where a new technology is being used that changes the very nature of food. Because it changes the nature of life, inevitably it will change humanity. So why shouldn't we proceed cautiously and expect public policies to be developed to make sure that public health is protected?

"This is an issue that has long-term implications for democracy. I think we're in a period where corporations have been virtually unrestrained in their activity in the marketplace, and we're starting to see the effects of that. And because the economy has been favorable for many Americans, there really hasn't been an attempt to exert some discipline into the conduct of these corporations. I think we're on the verge of a new era when Americans are reawakening to the importance of government oversight.

"There's almost a sense of corporations as Big Brother. People have spent a lot of time worrying about government as Big Brother. But we're in an era when corporations serve as Big Brother, and they decide what is good for people and what is not. There are democratic values at stake here—whether or not people have any right to have a say about how their food is made, what's in it. I think that we're passing through this period when corporations are making the rules and the government is giving them the legitimacy to determine our products without question. We're venturing into the unknown. Are we just supposed to eat our genetically engineered veggies and like it?"

Kucinich had come to a conclusion I had already reached: This issue went far beyond food to span science, agriculture, the environment, biodiversity, economics, trade, and the relationship between North and South.

The dimensions became even clearer in Seattle after the tear gas lifted. The debate that had sprouted across Europe and was taking root in the United States was also about culture. In genetically modified food, many skeptics see not just the threat of the unknown but an invasion of the culture of their countries. What we saw in Seattle was a backlash against the broader forces of globalization, of which genetically modified food is a symptom. The chaos in the streets gave notice of the emergence of a postindustrial politics in which people empowered by the Internet will be increasingly represented by the advocacy organizations they join. Leading the demonstrations, like the crop sabotage in Europe and India, was the vanguard of a determined movement.

Seattle is a place where things happen and movements begin. Grunge rock in music began there and, in fashion, emaciated, strung-out chic. In the labor movement, some of the early great mobilizations began there. The International Workers of the World organized first in Seattle. Then, in the Seattle General Strike of 1919, sixty-thousand workers walked off their jobs in support of striking longshoremen. Now, the smashed glass at McDonald's, Starbucks, Niketown showed the flammability of antiglobalization sentiment. Seattle was so traumatized by the protests and the discovery a few weeks later of explosives headed to a suspected terrorist that the city canceled its millennial celebration.

In the future, the WTO will no doubt learn how to insulate itself from disorder. But the global court of trade and commerce will be forever changed, forced to shed its arrogance and bow to civil-society demands on such issues as food safety, agricultural subsidies, and electronic commerce. The lesson from Seattle is the arrival of a new politics in which genetically modified food is at once a crucible for change and a singularly potent issue.

COLLAPSE

Genetically modified food is challenging the structure of the European Community, established by the 1957 Treaty of Rome. For three years, I have watched member countries ignore dictates of the fifteen-member organization which, over four decades, developed a *de facto* policy on food safety and which repeatedly concluded that GMOs pose no threat. Political exigencies within a nation's borders sparked rebellion, as did protectionist impulses. I am persuaded, too, that intransigent nations resented what they regarded as America's cavalier disregard of risk. For many reasons, the nations of Europe have refused to surrender sovereignty, rendering the European Parliament a toothless tiger in the arena with GMOs.

Europeans can't even agree among themselves. In Seattle, the agreement of European trade ministers to a Biotechnology Working Group apparently had lasted only hours before environment ministers of five countries—the United Kingdom, France, Denmark, Italy, and Belgium—issued a statement of opposition.

It didn't matter. The protests in the streets combined with balky issues inside the Seattle Convention Center to scuttle all deals. Around the world, people watched the negotiations falter. An example: *Al-Ahram,* an English-language weekly in Cairo, topped its front page with a photo of a WTO placard in flames. The cutline read, "Seattle, a normally laid-back U.S. city on the Pacific, has been turned into a global battle zone. Ten years ago, almost to the day, capitalism's final triumph was proclaimed as the Berlin Wall came tumbling down. The optimism was short-lived."

In Italy, columnist Vittorio Zucconi wrote in *La Repubblica* of the "strange but formidable alliance between environmentalist agitators and European ambassadors, between bluejeans and double-breasted suits, of mothers against Frankenfood—genetically modified food—and agricultural interests of Europe."

Three days into the talks, it became clear that the millennium would arrive without agreement to proceed with a Millennium Round of trade negotiations. With a mistakenly issued delegate's identification badge, I filed unrecognized into a closed-door committee-of-the-whole gathering before organizers publicly declared failure. A good deal of grumbling went on in the windowless room, as did a bit of soul-searching. Mike Moore, the New Zealander who held the post of director general of the WTO, told trade ministers bluntly that they need to reconsider how they operate. "We have to think very deeply about our culture. There is an inherent conflict in how we go about our business," he said, adding, "We have a deep institutional problem."

In the early-morning hours after the breakdown became official, anti-WTO tacticians distributed news releases that read, "Ding Dong, the Talks are Dead," and ministers tried to explain at a news conference what had gone wrong. Asked about the Europeans' fleeting consent to set up a special WTO Biotechnology Working Group, Pascal Lamy, the European Commission trade minister, divulged that Europe's tentative agreement had been nothing more than a bargaining ploy. "We only said it to help make a deal, to get something else. It is dead now," he said, speaking in French and presaging future confrontations between Europe and the United States.

Later, Brent Blackwelder, president of Friends of the Earth, climbed down from the soapbox—actually, a chair—where he had been trumpeting the success of environmental advocates. That day, his global organization had presented one of its "Earth-Wrecker" awards to Monsanto. "GMOs are the biggest mobilizing tool in the world right now," he crowed.

After the collapse, the forum for global decisions on genetically modified food shifted to Montreal as policy makers played a fumbling game of hot potato with this powerful new technology. Trying to make sense of Seattle for a newspaper story, I found notes I had written while talking to eighty-three-year-old Ruth Hunter during a protest outside the Seattle Kingdome. The Kingdome was an antiquated structure destined to be turned into rubble, but I resisted the urge to deploy that metaphor with the WTO.

Neither Ruth Hunter, who stood under five feet tall, nor any of her four California friends had resembled what they called themselves: the Dangerous Ladies of Santa Cruz. She didn't look like a threat, but she talked like a prophet. A prediction made by the feisty octogenarian stood out among words I had scribbled in the rain. "If anything will unify people, if anything will bring people together, it will be food," she said.

21

MONTREAL: LAST STOP ON THE GMO TRAIL

*Societies everywhere are perhaps now in a position to renegotiate the
technological covenants of how they produce food. . . . The bargaining
will not be easy.*
—Jack Doyle, *Altered Harvest*, 1985.

*Never doubt that a small group of thoughtful, committed citizens can
change the world. Indeed, it is the only thing that ever has.*
—Margaret Mead

IN THE MEN'S room of the Delta Hotel, I am standing behind John Herity,
a husky, bearded, Canadian diplomat who has suffered an indignity that I
am attempting, diplomatically, to measure. From his collar John Herity is
wiping away pie that moments before had been spattered in his face by a
wiry little man who disappeared faster than you can say, Resumed Session
of the First Extraordinary Meeting of the Conference of the Parties for
the Adoption of the Protocol on Biosafety to the Convention on Biological Diversity. By now, the pietosser is a block away from this ponderously
named but fateful gathering, catching his breath after bounding up two
flights from the subterranean ballroom, where the biosafety discussions had
commenced, then speeding through a revolving door into the snowy
morning. He's sprinted to safety and, by now, he's laughing, unlike Mr.
Herity, who is scowling into the mirror and dabbing at himself with a
paper towel.

"I don't know what kind of message they think they are sending."

Herity is speaking to a fellow from the biotech industry, who has appeared in the men's room to console him and who is professing his own disgust at "those people."

My motive for standing near a man wiping pie from a shirt that he may need to wear for the next twenty hours or so—if these negotiations go down to the wire—is, journalistically, straightforward: I want his reaction to being pied.

"What KIND OF PIE?" The words issue from Mr. Herity, who is repeating what I have asked as he squints in the mirror no doubt wondering who had asked something so . . . impertinent.

There are just two of us in this room now, and he sees that I am not unlike him, at least I'm not one of *those people,* dressed, as I am, in an olive-wool suit, and a shirt the same hue of blue as his.

"Cream pie," he says, softly. "I just don't want it on my shirt."

"I don't see any more," I say reassuringly, before noticing a dollop on the back of the collar of his gray, glen plaid jacket in a place he can't see in the mirror.

"Is it meringue? What kind of cream pie is it?" I ask.

Why this was important would be as unclear to me afterward as it was to both of us then. But John Herity took my question, as they say in diplomatic parlance, showing the amicability that had begun taking hold in these talks: There was a willingness to compromise afoot, even in the men's room. It was a spirit that, against the odds, might propel these delegates in Canada down a historic path toward resolving one of the era's great public-policy debates.

In a concession to me, Mr. Herity smacks his lips. "Just whipped cream," he says.

GLOBAL IMPASSE

Montreal in the winter is about as far as you can get from the Colombian seacost. But after the first round of biosafety negotiations unraveled in Cartagena, no government had stepped forth to bear the expense of playing host and, quite possibly, the ignominy of providing the dateline for news accounts of yet another failure. So the United Nations Environmental Program brought the negotiations back to Montreal, seat of early sessions of these five-year proceedings, and where, in the days before the formalities commenced, the temperature had not climbed higher than ten degrees below zero.

At the least, the frigid Canadian air would put a chill on the rowdies who had romped through the World Trade Organization meeting in Seattle seven weeks prior. As I'd heard it put, *those know-nothings will freeze their fucking asses off.* That comment was made by a coatless biotech representative strolling the quarter mile of underground passageways from the Intercontinental Hotel, where I, too, was camped and where their Global Industry Coalition convened in a war room at eight o'clock every morning, to the Delta Hotel, where most of the negotiating took place.

The industry people were right about the protests—but for the wrong reasons. Besides the pie aimed at John Herity and another at Joyce Groote, the blonde who heads the Biotech Canada trade organization, there were to be no Seattle-styled confrontations during the weeklong gathering in Montreal.

It wasn't the withering cold. Two days before the talks began, amid a windchill of minus forty below, over a thousand people rallied in the streets outside the International Civil Aviation Reorganization Building, where Montreal's United Nations operation is housed. On the day the formal negotiations began, arriving delegates were greeted by surely the world's most imposing ear of corn, a forty-foot-high omnivorous hybrid feasting on a monarch butterfly portrayed by a dangling Greenpeace activist. Near the surprising climax of the negotiations, several dozens of activists camped on the snow in a white-canvas tent, raucously chanting with derisive syncopation, *Shame on the Miami Group. Shame on Canada. Shame on the USA.*

But this was theater, not the hard marching in Seattle, where protesters had blared outrage and defiance at the amorphous evil of globalization. In Seattle, protesters were left out in the cold. In Montreal, the advocates moved about the proceedings purposefully but respectfully. The United Nations opened plenary sessions and even social gatherings to virtually anyone who had bothered to request an ID, and people felt less inclined to disrupt events they were part of.

In Seattle, locked out of the World Trade Organization literally and figuratively, the advocacy groups wanted failure. The fledgling alliance of environmentalists, labor unionists, and perennial protesters in the forty-thousand-strong assemblage, allied by a potent new strain of global populism, wanted the WTO to collapse and to go down spectacularly. But in Montreal, the environmental organizations gathered from across the United States and Europe, from as far away as China and Malaysia, hoped for success and to be on hand when the protocol was written. They saw no utility in locking arms to block delegates, as they had done in Seattle, or in scuttling the game.

The Arctic air had more effect on the delegates. When I walked into the United Nations building for the first time, still shivering myself, I watched African delegates fitted in army green parkas and zippered black boots courtesy of the Canadian government. For a week, delegates from the South looked like an invading army as they hiked across the street between the sterile U.N. headquarters and the Delta Hotel, where Colombian Juan Mayr, the charismatic chairman, wanted most of the negotiations. "There's a psychology here," Michael Williams, the United Nations press secretary, told me. "When people move back and forth between buildings, it's like moving from one world to another."

Rather than a vast hall with leaders on the stage facing the audience, Mayr ordered the plenary sessions configured in the so-called Vienna setting, where competing blocs sat at tables arranged octagonally, in the center of the six hundred delegates. None of the usual muttering from the dais and staring at a blurry fixed point beyond the audience. Negotiators from the United States–led Miami Group, the European Union, and the other blocs had to face one another in plenary sessions. That would have been impossible in the United Nations building across the street. Mayr told me early on he believed that the frigid climate could be a key to success. "Here, nobody goes out. I am grateful for the winter and snow."

After traveling tens of thousand of miles and writing tens of thousands of words on these issues, I believed that a biosafety protocol would be valuable—and for reasons beyond the personal reward of getting a good story that, according to my calculations, would break on Saturday just in time for the Sunday *St. Louis Post-Dispatch* with its half-million-plus circulation. A successful biosafety protocol could go a long way toward melting the global impasse over genetically modified crops and thus carry the world toward the middle, away from the extremes: On one side, the biotechnology industry, awash in arrogance, had allied with the self-interested Big Farm lobby and scientists hungry for grants and recognition—none of whom would concede risks. On the other, a potent pro-environment force that largely refused to concede benefits had bitten with bulldog ferocity into a fresh issue that swelled its ranks with lively converts.

But who knew whether this time, in a different latitude and in a new century, the result would be different? After the collapse in Cartagena, I'd written that reviving the negotiations would be daunting "judging by the lines drawn by the United States in the Caribbean sand." I'd based my prediction partly on the view of Veit Koester, the gloomy Dane who was chairman of the ill-fated Cartagena gathering.

I knew, too, that since Cartagena, the debate had evolved. There were, as the diplomats like to say, new facts on the ground: Top United States government officials had begun to question a strategy that had isolated North Americans from the rest of the world. The corporate life-science concept had begun to crumble. Mighty Monsanto, the ringleader of the technology, had been forced into a merger. European opposition was not only holding firm but had taken root in Japan and Australia, which were demanding labeling of modified products, and in Brazil, which refused to allow commercial planting of genetically engineered soybeans. American farmers had found themselves on the horns of a dilemma, lured to a technology that makes their lives simpler but frightened by consumer fears and diminishing exports. Like everyone else, farmers wanted a sign—and they wanted it before ordering their seeds for spring planting.

Just before the Montreal gathering, David Sandalow, the United States chief negotiator, had been stunned at the sentiments he'd encountered during a trip to Europe. "I've never seen an issue where the public profile is more different on each side of the Atlantic. That fundamentally affects the dynamics of this negotiation," he told me by telephone before the negotiations commenced. Sandalow was not optimistic about a breakthrough. "Quite the contrary," he said afterward.

With the backdrop of Mad Cow Disease and attitudes toward food starkly different from Americans', European consumers had grown no warmer to GMOs. As the sentiments in Europe stiffened, European delegates in Canada needed to appear resolute in their opposition. In Washington, trade officials looking on with alarm as export markets evaporated had cheered the European Union's effort in the preceding months to create a continentwide food-safety agency. Such a bureaucracy, modeled after the Food and Drug Administration, was proffered as the best hope for restoring faith in Europe's capacity to protect public health. But in Paris, two weeks before the biosafety negotiations opened, European leaders had unveiled a paper tiger. Their continentwide creation had no power even to inspect foodstuffs, let alone seize them or punish violators.

In Europe on the eve of the negotiations, the complex issues surrounding genetic science had been reduced to sloganeering: "better safe than sorry"—a reference to the "precautionary principle," which holds, in essence, that the absence of scientific certainty of harm is no barrier to protecting human health and the environment; and ratcheted-up "gate-to-plate" demands to label genetically modified products from before they sprout until the time they're eaten.

Margot Wallstrom, of Switzerland, the European Union's environment commissioner, had proclaimed the Europeans' commitment to these goals, and framed them as antithetical to the aims of the United States and Canada. Yet Wallstrom also had said that finishing the biosafety protocol was an "absolute priority," thus, seemingly, opening the door to compromise.

As the conference began, the Canadian press took a fresh look at the global ruckus and, as usual, it was not the sort of press that the biotech industry relished. Results of a survey that ran under the headline "Canadians Wary of Genetically Altered Foods" showed that two-thirds of people surveyed in Canada, the United States, and leading industrial nations would be less likely to purchase a food product if they knew that it had been genetically modified or contained gene-altered ingredients. Canadian papers reported farmers' fears and chronicled the protesters' every movement as they gathered for the talks. If delegates missed the stories, when they arrived they were handed similar clips from around the world, among them a story from Australia quoting a prominent business analyst named Robert Gottliebsen.

Farmers, retailers, or food processors "stupid enough" to ignore the consumer backlash against GMOs deserved the consequences, he said. "You can't come along and say 'here it is, our genetically modified food, put it in your lunch today and like it.' They won't cop it. You will get clobbered," he said.

In the news, negotiators saw the biotech industry clobbered without mercy. Frito-Lay, a division of PepsiCo Incorporated, had just announced it would no longer purchase modified corn for its chips. And there were rumors that Canadian-based Seagram's, one of the world's largest distillers, soon would reject modified grains. We were days away from America's Super Bowl football extravaganza, and it occurred to me that by next season, the chips and liquor that fans consumed in front of their televisions might not be derived from modified ingredients. In Montreal's *Gazette,* there was a story about the excesses of biomedical research, reporting that Canadian officials had spent a million dollars to house a monkey colony in homes with hammocks, toys, and natural trees. Then there was my favorite—a research report from Ottawa that moose milk might be the solution for people afflicted with lactose intolerance. Unfortunately for researchers, the cow gene they'd need to conduct their promising experiments would cost hundreds of thousands of dollars—because it had been patented. Patenting life-forms was, of course, a least popular bioengineering business method.

"There is really a nightmare," a scientist from Laval University lamented in the news story "and it's coming from the United States." The moral of the moose story: The biotech imperialists in United States of America were standing in the way of a cure for the world's gas.

What was needed was a valve to release the pressure. It was clear to me that the tide of strong emotions on this issue would not dissipate anytime soon absent a change in the status quo. The State Department had dispatched to Montreal the smooth, media-friendly Sandalow and a crew with a more felicitous manner than the hard-edged bunch that handled American business in Cartagena. But gracious negotiators did not mean the United States would relent.

I decided to have a chat with Veit Koester, the Dane who had chaired the failed Cartagena talks.

EX-CHAIRMAN'S LAMENT

I knew Koester had been irritated in Cartagena at both the intransigence of the United States and the manner of its negotiators. I'd heard whispers that he had been devastated by his failure to piece together a protocol that he had felt, at age sixty-five, would crown many years of work in global environmental politics. When we sat down in the Delta Hotel, I learned that the rumors were true.

"In Cartagena, I didn't know how the United States was until the very end. I must say, I miscalculated the situation. Always in the end, people come to some kind of agreement. It's extremely rare that you do not come to an agreement."

Koester said that when he had become a candidate to chair the biosafety talks in Cartagena, he had decided that it would be his swan song after working for years to achieve the Biodiversity Convention and other global agreements. And what a grand success it would be, bringing countries together on one of the most divisive issues in the world. But it was not to be, and, Koester confessed, it had hurt him deeply. "Personally, this was a big disappointment," he said. "I thought this should be my final chairmanship and I should be the one who made the biosafety protocol. It took me two months to get over it. I suffered for two months. Maybe not for two months, but let's say for six weeks, I negotiated every night; every night I had nightmares about it. It was terrible."

Koester had a decidedly more positive view of the U.S. negotiating team in Montreal after leaving Cartagena embittered at the Miami Group's tactics. "I would say that the United States delegation this time definitely is more civilized. They are not shouting so that everybody can hear them. They are not telling of their misgivings, showing it with their body language. In Cartagena, they were not civilized. I can't understand it, how it was possible for the United States to send people like that. Because they were not behaving as diplomats, with normal politeness and courtesy.

"But that is not the real point. The real point is that the Miami Group came to Cartagena with one purpose, and that was to block. You could link it to the WTO. In Cartagena, they knew about the upcoming WTO. They had in their minds there that they would try to move the whole issue from biodiversity to the WTO by proposing to have the working groups. If they had succeeded in doing that, then they could have come here, and said, 'No, no, no. We can't do this now; everybody agreed to a working group on biosafety at WTO.' Now, there's been this failure at WTO, and they realize that the plan they had did not succeed," he said.

In Koester's eyes, the stakes were even higher now because of the widening gulfs over modified food. "You cannot wait another couple of years," he said, "because it will be total anarchy with biotechnological products."

GLOBAL UNCERTAINTY

At a packed reception the evening the biosafety talks opened, activists and biotech representatives schmoozed with delegates. I made the acquaintance of two odd characters: "Frankentony" and a member of Greenpeace China.

Frankentony jarred loose breakfast-table memories from long ago; I was at a loss for words when an environmentalist acquaintance introduced me to this furry-suited, green-and-orange character who resulted, I was told, from a recombinant DNA experiment involving Frankenstein and the Kellogg Company's Tony the Tiger. Mary Shelley's character, now approaching two-hundred years old, still lurked near the surface of the genetic-engineering debate. But this Greenpeace reincarnation looked more like a lovable cartoon character, albeit one with a deranged look about him, as he passed out five-by-seven cards bearing his likeness and spoofing cornflakes boxes with an added enticement: "Hey Kids, get your gene splicer! See side panel for details."

Just as odd a hybrid was Lo Sze Ping. I'd written about China's vast fields of genetically engineered crops, first tobacco and now cotton. I had talked with the former State Department official hired by Monsanto to cushion acceptance of biotechnology in the world's most populous nation. So I knew China was buying into biotech and perhaps getting its money's worth. I'd heard from other sources that cotton farmers there had reduced their average applications of insecticide from twelve per season to about three, which was great for farmers and their land. But much about China's adoption of the technology was considered a "state secret," including the true acreage of modified crops, which the industry no longer included in its global estimates or, if they guessed at it, applied an asterisk. I also knew about China's head-cracking, murderous ways with protesters. So I was surprised to meet Lo Sze Ping, who handed me a business card identifying himself as a member of Greenpeace China.

When I sat with Sze, he told me that he was twenty-six and that he had studied anthropology. He said he lived in Hong Kong rather than on the Chinese mainland and that he and the nine Greenpeace campaigners in his office did not actually protest in the streets, nor did they uproot crops as had some of his European counterparts. But they would visit the mainland and make known their dissent, he said, and would stop in at government offices.

"The problem really, is that we really don't have a good documentation system. It is not really conveniently accessible to citizens to look at. China is not like Western civil society with freedom of speech. In the U.S., you can just browse in the Internet and go to web sites and find thousands and thousands of field trials, whatever you like. However, in countries like China, Monsanto can just, how do you say it, do something that is not accessible to the public. That's dangerous for farmers, who don't know what they are doing," he said.

Unlike some of the Greenpeace activists I had met elsewhere, Sze spoke cautiously and often respectfully about the government he wanted to influence. Chinese officials, he said, are astute about science and cautious when it comes to putting genetically engineered crops into mass food production. "From the government's point of view, if you don't know the risk, you don't want to fuck up, as we say; you don't want to create an instability in the society by messing around with the food system," he said.

Nor was Sze willing to totally dismiss the potential of biotechnology. "You cannot exclude a technology. But we feel it has nothing to do with improving people's lives," he said.

"China is maybe the biggest agricultural production system in the world. There are thousands and thousands of experts on food and agriculture. These people are not dumb. They will not buy a very simple equation promoted by Monsanto that if you use Roundup Ready, you can feed your people. They have been feeding 1.2 billion people, and they don't need Monsanto to tell them how to do it," he said.

Buttonholing delegates from other developing countries, I found colliding sentiments. Cho Jai-Chul, who headed the South Korea delegation, told me he believed that a biosafety protocol would serve to promote an industry with promise. Maybe, he said, fungus resistance could one day be engineered into the country's hot pepper crop, which was vulnerable to disease. "You know, we like a lot of, how do you say, spicy food," he said.

Burkina Faso, a West African country that I had forgotten existed, sent delegate Kambu Jean Baptiste to look out for its interests in the negotiations. Baptiste, like Jai-Chul, said he was hopeful about the potential of GMOs to help his country. "We have lands where nothing is growing and people have no food, or medicine, and people take plants to get well. It is a dangerous life, and maybe GMOs can help," he said.

I ran into Juan de Castro who, more than a year before in Brazil, had set me thinking when he noted that in this technology the North, not the South, was the guinea pig. At the time, de Castro, a geneticist, was heading Brazil's biosafety commission and had to tread a middle ground, publicly at least. Now, working for a government research agency, he was free to trumpet what he saw as biotechnology's benefits to Brazil. "In order to be able to compete internationally and sell products, we need to be able to use biotechnology to reduce costs, particularly chemical costs," he said. "Brazil has a strong tradition in terms of genetics for the tropics.

We need genes to make plants capable of withstanding adverse conditions: aluminum toxicity and drought. These will come with biotechnology. Pretty soon, we will be able to identify genes that will make plants more efficient."

In Cartagena, however, Brazil and China were aligned with most of the rest of the world in the Like-Minded Group, the alliance that had held out for a strong protocol rather than no protocol at all. In these countries, many were suspicious both of the technology and of the multinationals delivering it. They worried about theft of their genetic resources, of their knowledge, indeed of their body parts, by bioprospectors. But they didn't want to be cut out of the bounty that might come from biotechnology. So they hung together against the United States, usually siding with the European Union, in demanding a tough-worded treaty.

The European leaders I spoke with before bargaining began signaled no inclination to give ground. Meanwhile, the United States–led Miami Group remained intact, although I couldn't be sure of its resolve. New to the negotiations, thanks to the efforts of Juan Mayr, was the Compromise Group, which was led, fittingly, by Switzerland.

At stake, once again, were rules that would govern the movement of billions of dollars' worth of genetically modified products. Like K Street lobbyists outside the Gucci Gulch hearing rooms of Congress, the cream of the industry strategists and trade associations had trekked to Montreal to fight against words that could cost them real money and, worse to some of them, freedom to do business. At the reception, beleaguered Monsanto officials wore their pink identification badges backwards so people wouldn't know who they were.

With so much at stake, I knew that it would take uncommon talent to engineer a compromise. In Colombia, I had met Juan Mayr Maldonado, who had taken over for Veit Koester as chairman of the proceedings. Unlike Koester, a lawyer who seemed oppressed by the weight of his task, Mayr was jovial, effusive. He was tall, a shade over six feet, with flopping brown hair parted in the middle, a shambling gait and time, maybe even a hug, for everyone he encountered. Mayr was a photographer by trade, and I'd heard that he had polished his negotiating skills atop an isolated mountain mediating disputes between drug cowboys, revolutionaries, and Indian tribes. The parties to these biosafety negotiations might be no less intractable.

NEW CHAIRMAN'S HOPE

After watching Juan Mayr operate for two days, I raised my odds of success in these negotiations to fifty-fifty—still more favorable than I was hearing from negotiators or advocates from any group. When he joined me for breakfast one morning, I asked him to tell me about his role in quelling the Colombian uprisings.

On a napkin he drew a mountain, the Sierra Nevada de Santa Marta, which, he said, was the highest coastal mountain in the world, nearly six thousand meters. It was the region of the Tairona people, he began, relating some of their history, beginning with the arrival of the Spanish in the 1500s.

"They started to take the most fertile lands of great valleys and put in all these settlements. And when you take the forests out, you immediately have erosion. After long fighting, the indigenous people lost their last land," he said.

In the 1950s, victims of political oppression elsewhere in Colombia started moving to the mountain, and many of them planted coffee. Then came the marijuana growers with their chemicals and their weapons. "A great boom of marijuana started," he said, providing more of the background of the challenge he faced. "The forests started to be destroyed, more than a hundred thousand hectares. And that immediately affects the water because this mountain, with its thirty-two rivers, is a great water factory. So it started to affect all the surroundings. All the marijuana caused the indigenous people to be pushed up and up and up the mountain, where the quality of land is not good and where they need more time to produce food," he said.

In the 1970s, Mayr arrived. He set up an environmental foundation and in so doing picked up experience in mediating among indigenous people, the political refugees, and the pot growers.

The stew of people at Sierra Nevada de Santa Marta began to boil in the 1980s when the guerrillas arrived, first one faction, then another, and then a third. Then the government paramilitary forces began showing up to fight the guerrillas. Mayr, environmentalist and photographer, became a peacemaker.

"What I had to do, I made approaches to all of these groups, the indigenous people and the farmers and the marijuana people and the political people and the guerrillas and the paramilitary, to try to give them information so everybody can share a common vision. We had workshops but first we had to work informally with one group, and then another, trying to build confidence, looking for the conservation of this mountain. It's been a long process, a very, very long process," he said.

It sounded to me like the preparations leading up to this gathering, and I asked Mayr if there was any comparison to his negotiating task in Montreal. "The biosafety talks have some similarities," he said. "It is a complex equation, a very complex equation."

After eating breakfast with Mayr and hearing about his world, I phoned an editor to say that I thought that a deal would happen.

PRECAUTIONARY TALES

By the third day, a global agreement on biotechnology seemed too complex for even Juan Mayr's high-altitude skills. Canada, the United States steadfast ally, had not yet ranked the negotiations high enough to dispatch federal Environmental Minister David Anderson from Ottawa. Activists had posted yellow wanted posters with his bearded face on bulletin boards in the two buildings where negotiators gathered.

A sub-group of negotiators had agreed tentatively that the protocol would cover commodities, chiefly genetically modified grain shipments from the United States. That was, in my estimation, a step forward since Cartagena, where the Miami Group barely would entertain that notion. Now the full scope of coverage had to be fought through. The Europeans demanded a world trading system in which shipments of GMOs would

be identified by their genetic makeup. That would require extensive testing and, most likely, segregation of modified crops. That was a chilling
prospect to grain companies, who would need to figure how to spread
out these costs, and to American biotechnology companies, who had continued to insist that gene-altered varieties were no different than conventionally grown varieties.

The disagreement over labeling commodities was a huge barrier and,
quite possibly, a deal-buster. So was a long-festering dispute over the precautionary principle and the Europeans' insistence that it be part of the
protocol.

In the Delta Hotel's ballroom, leaders of negotiating groups hinted
grimly at their differences. Kristof Bail, the German who represented the
European Union, spoke passionately long into the night about the commitment on his continent to the precautionary principle, which was anath-
ema to American companies. Precaution had long been a condition of
regulation in Europe, back to the nineteenth century, when British doctors
had struggled to find the causes for cholera in the polluted rivers around
London, Bail said. "We need to have some guidance, all of us collectively,
as to what should be the yardstick for governments faced with environmental uncertainty," he said.

The modern precautionary principle was born in West Germany's water protection law in 1972 and spread across Europe. It was swiftly incorporated in environmental treaties starting in the 1980s and became known
globally after it was written into the benchmark Rio Declaration on Environment and Development, which the world adopted in 1992. The
principle holds that governments should have the means of protecting
public health and the environment even in the absence of clear, scientific
evidence of harm. The debate over precaution has an element of seman-
tics, but there are nonetheless clearly different approaches toward safety in
Europe and the United States.

In Europe, where Mad Cow Disease had ravaged faith in government
and scientists alike, regulators assert their right to protect people from the
unknown. In the case of genetically modified food, that means preventing
products from reaching the market.

In the United States, the philosophy that's evolved in the last quarter of the twentieth century is a quantitative, risk-based approach to regulating public health and the environment. After calculations of risks and bene-fits—which in the United States often are left up to corporate science overseen by regulatory agencies—judgments are rendered as to whether the value of a product or a technology is worth the risk of commerciali-zation. In the biotechnology debate, these different approaches have been reduced to competing slogans: Europe's *Better Safe Than Sorry* was com-peting against the United States mantra of *Sound Science*.

Juan Mayr persuaded people to examine their most tenaciously held beliefs. He lightened moments with humor and hugs. He passed out teddy bears, bringing smiles to the most hard-bitten negotiators. To weary del-egates stumped at how to resolve different approaches to precaution, Juan Mayr sounded more like a Latin American novelist than a United Nations functionary when he concluded the late-night plenary session.

"I want you to have good dreams tonight with the precautionary ap-proach so tomorrow we can share our good dreams," he said.

A MODIFIED MEAL

At dinner that night, environmental ministers could not help but look suspiciously at their plates. They had gathered at the St. James Club in downtown Montreal for a sumptuous spread whose contents had been leaked to Friends of the Earth. When the ministers, negotiators, and guests arrived at the club to dine, the activists greeted them, passing out anno-tated menus:

CREAM OF SWEET POTATO SOUP
A GM herbicide-tolerant U.S. specialty with a touch of special GM corn starch thickener.

ST. JAMES SALAD WITH VINAIGRETTE
Some delicious GM virus-resistant peppers served on a bed of herbicide-tolerant lettuce and lightly drizzled with a variety of GM oils, including our popular corn and soy oil. And, of course, not forgetting the GM tomatoes.

GRILLED SALMON WITH WHITE WINE CREAM SAUCE
Why not try the fast-growing GM fish variety? Made with cream from GMO-fed cattle. Coming soon, specialty GM wines—French or German?

FILET OF BEEF WITH THREE-PEPPER SAUCE
Juicy meat from GMO-fed beef cooked to your own taste.

ENTREES SERVED WITH SEASONAL VEGETABLES AND POTATO
Current selection might include herbicide-resistant cauliflower, broccoli and peas. Not forgetting a range of GM potatoes from around the world.

CHOCOLATE CAKE
Almost certainly includes eggs from GMO-fed chickens. Lecithin likely to follow from GM soybeans.

COLOMBIAN COFFEE, THE BEST IN THE WORLD.
So far not GM, and let's keep it that way.

"CRITICAL JUNCTURE"

The arrival of ten European environmental ministers renewed hope that an agreement would be forged in Montreal. But the Miami Group read their arrival as a mixed message: On the one hand, they brought a new flow of high-level energy; on the other, a new cadre of politically sensitive Europeans would have to be dealt with. To beef up the American negotiating team, Undersecretary of State Frank E. Loy arrived. Loy struck me as sensible of both the European psyche and the exigency of cutting a deal. He was back in the world of diplomacy now, but for fourteen years he had run the German Marshall Fund, which focuses on matters affecting both Europe and the United States. I knew, too, that he had served a spell as chairman of the Environmental Defense Fund, a New York–based advocacy group that had registered skepticism in the debate over genetically modified food.

On the next-to-last day of negotiations, the European environmental ministers put out a news release pronouncing the talks at a "critical juncture" and restated to reporters their commitment to the precautionary principle. "We do not know everything about the long-term effects of GMOs, and we have to be able to use the precautionary principle," Margot Wallstrom of Switzerland, the European Union's environmental minister, told us.

There were more sticking points: Europe insisted that a protocol not be subordinate to other international agreements. By other agreements, the Europeans meant the World Trade Organization, where trade disputes over GMOs seemed sure to land. They would not, they insisted, permit a protocol to be undermined by the WTO.

The final demand reached to the core of the global debate over modified foods: labeling. To the dismay of the Americans at the negotiations, most of the world continued to demand a system of documentation that would let people know from "gate to plate," as the Europeans liked to say, if their food had been genetically engineered. The Miami Group already had rebuffed a drive to require prior consent to shipments of modified grain. A late version of the protocol had words that riled U.S. agribusiness and backed the Miami Group against the wall.

The plan on the table would, as best I could tell, require testing of commodity shipments—corn, soybeans, and probably much more—along the way in order to determine their genetic composition. Importing countries would have the documentation to know what type of genetic engineering had been used and by which company. The Miami Group proposed a new clearinghouse for information that, it claimed, would give the world sufficient information on safety and science. But the Europeans insisted on more. Requiring documentation with shipments, they insisted in a news release, "is of high importance to the credibility of the protocol. The protocol will simply not work if the importer cannot verify if a shipment is in conformity with its provisions. All parties must know what is in a shipment in case of a spill."

That proposal would put the world on a fast track to a two-track food system, one modified and the other conventional, with the so-called natural identity of commodities preserved. Food distributors didn't want such tracking because the costs could be astronomical, they claimed. Biotechnology companies didn't want it because it could threaten the future of their products.

The dispute over documentation that exploded in Montreal was part of the larger debate over mandatory labeling. Bare-all labeling might be inevitable one day at every step of the way in food production, but not in Montreal, prayed the Miami Group and the industry representatives waiting nervously outside the negotiating room.

The *New York Times*'s Andy Pollack and I—the only reporters from American dailies who had covered both the Cartagena and Montreal proceedings—wrote stories for the next day about this threat to the potential success of the gathering. Each of us had filed versions with anecdotal first paragraphs to ease readers into a story about an obscure concept. His top remained but, mine, I saw when the story appeared in print, was edited out, so that the story began with what I had written as my second paragraph: *A tricky concept known as the precautionary principle has emerged as a key stumbling block to completing a treaty that would govern the movement of billions of dollars in genetically engineered products around the world.* (With a top like that, I worried that the tricky precautionary principle would be costing me my readers.) I quoted Frank Loy, who made it clear once again that America's worry had to do with trade. "There are some countries in which it will have the effect of justifying restrictions and even complete import bans. We have to worry about that and we do worry about that," he said.

Negotiations had reached an impasse. Given the two giant obstacles remaining—the clash over the precautionary principle and the dispute about documentation of shipments of GMOs—achieving a biosafety protocol seemed about as likely as a thaw in this Canadian January. As improbable as a rabbi smuggling drugs. Yet the morning of the final day of negotiations, Canadian papers forecast a rare warming trend. And I noticed a curious story about a Montreal rabbi sentenced to home detention for smuggling balloons packed with cocaine and marijuana to prison inmates.

HARD BARGAINING

Negotiations typically are grueling, unpredictable, and messy. The bigger the stakes, the longer they take, which is why I scheduled my flight home the day after they were supposed to conclude. I knew that Chairman Juan Mayr, who was fond of talking about dreams, had to be in dreamland when he predicted that the final negotiations would be wrapped up by midafternoon Friday, in time for everyone to visit Montreal's Biodome. There, in what looked like a giant flying saucer alighted in central Montreal, delegates would visit five separate ecosystems from around the world.

On Friday morning, after working until 5:45 A.M., sleeping for two and a half hours, and taking a cold shower, Mayr seemed less hopeful as we sat together over his minimalist breakfast of coffee, orange juice, and white toast with grape jelly. Nor, when the plenary session convened a short time later, was he so sanguine as the night before. He told delegates, "Right now, I have a fresh view, and I am very optimistic that today, we will come to full conclusion for a protocol. But it will take some time, and, I'm sorry to say, we will lose the opportunity to go to the Biodome."

Throughout that final day, the odds of reaching an agreement worsened. As negotiators shuttled between floors of the Delta Hotel, it looked as though the ever-cheery Mayr might be headed to the Biodome alone, to sulk. There was no evidence of progress, and Mayr already had postponed the concluding plenary session from afternoon to evening. Planes were being missed. In hallway conversations, participants already were planning damage control. If the agreement wasn't reached that night—if by midnight the First Extraordinary Meeting of the Conference of the Parties for the Adoption of the Protocol on Biosafety to the Convention on Biological Diversity wasn't called to order—the world would not have a biosafety protocol. Perhaps, talk had it, the protocol could be finished in Nairobi in May.

If ever I am kidnapped on foreign soil, I'll be heartened if the tenacious U.S. State Department negotiates my release. As ex-chairman Veit Koester had said about the United States team in Cartagena and as the Europeans grumbled in Montreal, the American negotiators are tough sons of bitches. As the three P.M. plenary had been, the eight o'clock wind-up was put off for lack of an agreement. That session now was set for eleven o'clock, literally the eleventh hour. But eleven passed like every other deadline in Montreal and Cartagena. The Europeans, whose body clocks by now were expecting sunrise, appeared deflated. In a hallway, a red-eyed Michael Meacher, Britain's environmental minister, said nervously that no agreement was at hand; negotiations could collapse at any time. Kristof Bail of Germany complained to me that the Miami Group will not move. "At Cartagena, we found at the end of the day the position of the United States and Canada was 'our brief or nothing.' At the end of this day, it could be the same thing," he said.

The State Department's Brooks Yeager had tried to explain the ramifications of a stringent system of documenting shipments. It would cost tens of billions of dollars, Yeager said in a meeting of Europeans and the Miami Group, and the Europeans seemed at least to be listening. Americans at the table were astonished by how little these European officials charged with protecting the environment knew about agriculture. Perky, blonde Margot Wallstrom, the European Commission's environmental minister, looked slightly out of place in her hiking boots among a roomful of black oxfords. Still, the Americans were happy she was there, seeing her as "extremely constructive." They weren't feeling any warmth toward Svend Auken, the hulking foreign minister of Denmark, who lowered the room's temperature with *ad hominem* attacks on the United States. "They haven't even ratified the Convention on Biological Diversity," he bellowed at one point, "and they're trying to dictate to everyone."

During a lull, I tracked down Auken, a six-foot-five giant with hair the color of snow. He was large and he was angry, and while we were talking, I pictured him losing control and flinging Sandalow and the other Americans ass-over-elbows out the meeting-room door. Auken has the bearing of a military officer or an actor; he looks like he could be anything he desires. What he really had wanted was the top job in Denmark, prime minister, but a political shift had frustrated those hopes. Auken has taken a special interest in these negotiations because, he told me, the first biosafety meeting in Europe took place in his hometown.

"They're arrogant," Auken said of the Americans. "They are blocking a protocol to a convention that they are not even party to. It's like Cartagena all over again. We don't want their polemics. We just want results. We just want them to let us do the consumers' business. It is not only being able to label. We should know what the products are that we get. But if they are bundled together, you can't do that. You can't give consumers a choice. It's impossible."

JUAN'S WAY

Juan Mayr felt success slipping away. He was about to lose the Cartagena Protocol, named after his seacoast city. In what looked to all like a last-gasp effort, he visited the United States delegation with a proposal. He said the Miami Group might get a concession, some vague wording, on documentation of shipments. But, he added, there was no hope of striking precautionary wording.

Mayr's visit prompted two consequences. First, the members of the Miami Group held their own impromptu gathering to debate prospects. Then members from key delegations got on the phone to their capitals. The U.S. delegates reached senior Clinton administration officials at home.

"Nobody thought it was very good; nobody thought it was ideal," the State Department's David Sandalow said. "The question was whether it was acceptable."

Mayr needed more time, so just before midnight he convened the plenary and deployed an old time-stopping ploy that I got to know as a fledgling reporter covering the Illinois General Assembly in Springfield. "I would like you all to stop your clocks," Mayr said.

When legislators from Chicago and downstate Illinois ordered the clock stopped just before midnight on June 30, they could squabble over money until the Fourth of July—or later. So in Montreal, in the freezing early hours of January 29, the negotiations continued timelessly. But with so many international flights looming, they wouldn't last beyond daybreak.

Outside, fifty or so protesters chanted *Shame on the Miami Group. Shame on Canada. Shame on the USA.* Freezing or not, we might yet get a Seattle-styled riot.

I ran into Tewolde Egziabher, the Ethiopian who, as head of the Like-Minded Group, spoke for China, India, and about four billion people in these negotiations. I knew where his sentiments lay—strongly suspicious of biotech—because I'd taken him to lunch in Cartagena. "We've done the best we can," he told me with the Montreal negotiations stalled. "Perhaps we should go our own way and pass our own rules, and then anybody who wants to trade with us has to play by our rules. And then the United States can come back to us in two years or so and tell us if they like it."

The biosafety negotiations were perched on the precipice of another failure. Delegates and industry reps were now worrying how it would look to the world. Willy de Greef of Novartis repeated to me his company's anxiety over another collapse. "This is no bullshit thing; this is self-interest on our part. We're not interested in the breakdown of these talks."

When it gets to be three-thirty in the morning, negotiations become Darwinian. The rich countries, who've been able to shuttle in fresh thinkers in clean shirts after their naps, beat down the weary. I'm among the down-beaten; I find a chair in which to sleep. The *New York Times*'s Andy Pollack is splayed out on the floor, still in his trench coat, looking like one of the homeless in the city that sends him his paychecks.

By the shuffling around me, I realize—with eyes still closed—that it's over. Negotiations have ended. There is agreement—tepid, middling, but still agreement—between the Miami Group and the European Union, the principal combatants here. But the rest of the world still has to ratify it, and given Tewolde's judgment, the rest of the world may be of a different mind. In Cartagena, the Like-Minded group wanted no part of a toothless agreement.

It is 4:30 A.M., almost rum-drinking time by Cartagena standards, and from the stage in the United Nations auditorium Tewolde beckons delegates to come to him. As delegates from every corner of the earth approach him this last time, the fate of these negotiations and the course of a new global politics of food is at stake. Leaders of all the negotiating groups had met with Juan Mayr minutes before and agreed, tentatively, to his proposal. But one diminishing word from Tewolde might bring it all down.

Then, anything could happen. Suspicions about genetically modified food would be reinforced and, rightly or wrongly, the halting advance of a new technology might cease, its promise never to be tested.

Tewolde spoke softly, reassuringly I thought, and without judgments. Yes, he told his delegates, the agreement did reference the precautionary principle. And from a single sheet of paper he read aloud the one paragraph of stilted wording that contained the compromise between the Miami

Group and the European Union on the final sticking point: how the world will handle the shipments of genetically modified commodities.

Any shipment containing "living modified organisms that are intended for direct use as food or feed or for processing" must carry the words "may contain living modified organisms" along with directions to more information. No further decisions regarding labeling would be made until two years after the protocol went into effect. In other words, any shipment of modified grain might contain vague notification on the bill of lading. But the hard decisions on testing, liability, and disclosure of precise genetic constituents of food had been put off.

The United States and its partners had swallowed the precautionary language along with the subjective powers it could give other nations to control shipments of modified products. But they won their concession: Only baby steps would be taken toward labeling.

"Shall I go and say we accept?" Tewolde asked.

A few grumbles sounded, but they were swiftly drowned by the words "Yes, Yes, YES."

"Good. We have a protocol. I think that this is the best we can do," he said matter-of-factly.

At 4:42 A.M., Juan Mayr convened the Resumed Session of the First Extraordinary Meeting of the Conference of the Parties to the Convention of Biological Diversity. What I saw in the predawn Canadian morning was extraordinary. When Juan Mayr proclaimed success, Monsanto officials and Greenpeace activists, the main antagonists in one of the most profound and acrimonious debates of the era, both stood to cheer. Many speeches sounded, but I transcribed only Juan Mayr.

"I love you all," he said.

"I congratulate you for your success. This represents a victory for the environment, for the international community, and for citizens throughout the world. This is just the beginning. It represents a great challenge."

PLANTINGS ELEVEN

Returning from a trip just before Labor Day weekend, I peered out my window. What I saw first was the dew on my Moon and Stars watermelons reflecting the early sun. My eyes shifted to the purple flowers of Grandpa Ott's Morning Glory that had covered the split-rail fence in my absence and climbed to the top of my bamboo tomato stakes.

But a glorious morning it was not. When I fixed my gaze between the melons and the morning glories to check on my soybeans, I saw . . . nothing. I moved outside to my deck and looked down, confused. It was no mistake. I knew that before I had climbed down the hill, barefoot, and stood in the barren swath of dirt. My experiment, my two rows of pirated, genetically engineered soybeans and my remaining row of conventional beans, were gone. Nothing remained but a few pods in the dirt.

Before leaving town, I had squatted between the rows and concluded that harvesttime was near. Both varieties had grown heartily, to about three feet, in dirt worked with compost and fish entrails. Their growth wasn't my doing; all the rain during spring and summer had blessed my region with bounties of all sorts. Already, my neighbor's tobacco hung in his barn to dry.

It had begun as a lark. Perhaps the engineered variety would sprout misshapen in some diabolical Frankenfood way, and I would write about it. As the beans grew, I became attached to them. The gene-altered bushes might have been a tad taller, but measurement time hadn't arrived. Their tough, veiny leaves—sprouted from seeds modified to resist weed killer—had stretched from wrist to second knuckle. Clusters of green pods, looking like fuzzy baby heads, had begun to fatten and dry on their cilia-covered stalks. They were majestic plants indeed; I understood why the Chinese called soybeans "Heaven's Bird." They were as wide as they were tall, each of them with fifty or sixty pods containing three beans each. So much protein in those egg-shaped little nubs. In my backyard, I had grown "the most important food in the world," and I was proud.

I'd planned to roast some of the beans and eat them as snacks. I'd run across a recipe for miso—fermented tofu—perfected by a sixth-century Buddhist monk.

But my beans were gone. They'd been disappeared, to borrow a verb usage common in the time of Guatemalan death squads. But who had done this? Who sabotaged my test plot? And why?

Recently, in Scotland, three saboteurs had been asked by a judge why they had attacked a test plot of modified canola. One young man said he wanted to kill the plants before they flowered, distributing pollen thereabouts. Another said he was convinced the herbicides sprayed on them were killing other plants. A third said only that he shopped at an organic farm a quarter mile away. Apparently he had wanted to protect the integrity of the natural produce.

I calculated. Was it deer? Unlikely, considering that they nibble and tear, not yank from the soil, as my denuded pepper plants nearby proved.

Was it my neighbors, some of whom hadn't approved of my experiment? Then there were the environmentalists down the road who, over the course of nearly four months, might have heard about what I was growing. What about my wife? One night, she had read aloud from a magazine quoting somebody saying that once you grow modified plants, your soil is forever changed.

I couldn't be sure, but I would find out.

AFTERWORD

In the spring of 2002, I found myself back where this book began—handicapping the future of an American-hatched technology with the chief spokesman for the American food industry in the uniquely American milieu of Major League Baseball.

Two years had passed since I had sat at Camden Yards in Baltimore with the Grocery Manufacturers of America's Gene Grabowski, taking stock of subtle alterations in the mosaic of life: the presence of genetically modified ingredients in chips, soft drinks, and a range of ballpark fare—as well as in fans' shirts, now woven from the bolls of gene-altered cotton.

I said back then that gene-altered agriculture probably was here to stay, but that public acceptance would take years, perhaps many years. Why would consumers knowingly bear the uncertainties of a technology that, for now, offered them so little? Gene, a former journalist, went so far as to predict a war over food, with skirmishes escalating between the alliance of the biotechnology and food industries on one side, and the worldwide resistance of activists, consumers, and researchers on the other. Ultimately, he had said, one side would win. "We'll either go biotech or we won't."

By 2002, Orioles' "Iron Man" Cal Ripken had retired, a testament to the fact that stories almost always have endings. But for genetically engineered food, many chapters had yet to be written.

There was indeed war in the world—not over food, but over terrorism in the aftermath of the September 11 attacks. As with many public policy issues, some of the urgency surrounding GMOs had become at least temporarily diminished.

In the third spring of the industry's messianic campaign to persuade us that the future belonged to genetically engineering, we were being told that a world made healthier by biotechnology grew ever closer. The latest tabulations showed that the global acreage of modified crops had climbed to 109.2 million acres, a twenty-five-fold increase over 1996.

More engineered soybeans were sprouting in the United States, Argentina, and, illegally, in Brazil. India had consented to commercializing insect-resistant cotton, in part because of fears in New Delhi of progress in China.

But by my reckoning, the food supply had remained remarkably resilient to the forces that would re-engineer it. Plantings of modified corn and canola had declined. Ninety-nine percent of the world's transgenic crops sprouted in just four countries and, as two years before, 68 percent had been sowed in the United States.

Beyond papayas, I could find no other genetically engineered whole food on the market. Unlike two years ago at the ballpark, there would be no modified french fries; Monsanto had taken bioengineered potatoes out of production. Nor, any time soon, would there be bread from modified wheat. The biotech companies continued testing wheat, government records showed. But farmers themselves feared the consequences of re-engineering the staff of life.

"Wheat is mentioned in the Lord's Prayer. Other commodities are not," Alan Tracy, president of United States Wheat Associates, said in explaining his industry's reluctance.

Sitting behind first base, I had the sense of being suspended in time. A new season of protest had begun after months when street resistance of any variety seemed unpatriotic in America. A week before, the Organic Consumers Association had staged corn-dumps and guerrilla theatrics in the United States, Canada, and Mexico.

In Europe, still the center of global resistance, saboteurs in Belgium had just destroyed a test crop of rapeseed, one of the few engineered crops being tested on the continent this season. The European Union had proposed new rules for labeling and tracking genetically modified products in hopes of assuaging consumers and resuming approvals for planting corn and other modified crops. But a political solution remained elusive.

Even the scandals sounded familiar. The admission by biotech companies two weeks earlier that unapproved canola may have entered the American food supply had raised anew fears that the industry was unable to control its brave new crops. So had the mystery surrounding the discovery of modified strains of corn in remote regions of Mexico, the international center for corn diversity.

But the biotech barons could take solace from the fact that Americans, unlike the Europeans, were paying little attention. Polls also showed that when people listened to the arguments surrounding GMOs, they were nearly evenly divided in their sentiments for and against.

The United States government persisted in its refusal to seriously consider the labeling of gene-altered foods. Nor had federal agencies responded to the admonition from the National Academies of Science that gene-altered crops needed more stringent regulation. But in their 2002 report, the scientists had found no proof that modified crops harmed the environment.

Meanwhile, the Cartagena Protocol on Biosafety that was so painfully completed in Montreal had received about one-third of the fifty ratifications it needed to take effect. Predictably, the United States had not blessed the pact that it had labored to weaken. On the way to becoming attorney general, then-Senator John Ashcroft had signaled opposition to the treaty with his barbed comments to then–Secretary of State Madeleine Albright. "I think it threatens very substantially the technical position of the United States and our capacity to feed a hungry world," he said in a Senate hearing.

At the ballpark, Gene Grabowski and I talked about the ceasefire in the food war, if that's what it was. Two years before, his cell phone's ringing had interrupted us; there were questions from reporters about one of the GMO dustups in the news. This time, the phone was still.

"The war may not have been won, but it has been suspended," he said.

My sense was that passions would become inflamed again. In the coming months, the Food and Drug Administration would be weighing the commercialization of transgenic salmon. The industry also was conducting experiments aimed at harnessing plants to produce pharmaceuticals, chemicals, and even plastics—a deployment of the technology that would arouse new concerns.

For now, as players in both baseball and biotech took the field for a new season, the spokesman for the food industry displayed a victor's confidence. "Want some genetically modified Crackerjacks?" he joked.

I'm still not sure who destroyed my engineered soybeans. The activists I suspected moved away before I could corner them. And neither my wife nor any neighbor confessed. A fast-growing magnolia blocks the sun where they sprouted and I have no more experiments on tap.

But I hear that Emelia, an eleven-year-old in my neighborhood, is conducting her own experiment with gene-altered seeds for a science project. That is fitting, for it may be her generation that decides their fate.

—BILL LAMBRECHT
April 2002

INDEX

- 3 pages on PROS/CONS for farmers across the World
- 3 pages on PROS/CONS for public all across the world
- 1 page on POSITION + why you take that position + why + which sources of persuasion + which arguments influenced you the most.

Movie Notes

papaya lawsuit - GM papaya + Intellectual rights

FARMERS	PUBLIC
PRO	PRO

FARMERS

PRO
1. ease of farming p.100
   GM seeds protect the plant (pg. 104)
2. Profit ← evidence(41)
3. vanity

CON "seed and chemical"
1. intell. rights + ownership corporate (p.142)
2.
3.

CON
1.
2. "seed/chemical %(p.142-3) corporate ownership of land + farmers" p.152
3.

PUBLIC

PRO
①
②
③

CON -SAFETY-
pg.9 ① health risks unknown (76)
② control of food in hands of few people
③ New tech w/ power to reorder the building blocks of life.
evidence: cry 9/other countries not going along w/it
p.43
p.142